Division Facts

in five minutes a day!

Susan C. Anthony

Instructional Resources Company
P.O. Box 111704
Anchorage, AK 99511-1704

Teach with less effort, more SUCCESS!

Instructional Resources Company is a teacher-owned business which aims to assist teachers in maintaining high objective standards while helping *all* students reach them. Our materials and workshops are designed to save teachers time while helping them nurture the excitement of learning in their students, build confidence through early success, and build a framework of background information to which new learning can be connected throughout life. We welcome feedback from anyone using our materials with students!

Instructional Resources Company
P.O. Box 111704
Anchorage, AK 99511-1704
(907) 345-6689

Instructional Resources Company
3235 Garland
Wheat Ridge, CO 80033

The following materials by Susan C. Anthony are also available. Most are reproducible.

Mathematics	*Addition Facts in Five Minutes a Day*
	Subtraction Facts in Five Minutes a Day
	Multiplication Facts in Five Minutes a Day
	Division Facts in Five Minutes a Day
	Casting Nines: A Quick Check for Math Computation
Reference	*Facts Plus: An Almanac of Essential Information*
	Facts Plus Activity Book
	Encyclopedia Activity for use with The World Book Encyclopedia
Spelling	*Spelling Plus: 1000 Words toward Spelling Success*
	Dictation Resource Book for use with Spelling Plus
	Homophones Resource Book
	Personal Dictionary Masters
Handwriting	*Manuscript Handwriting Masters*
	Cursive Handwriting Masters

ISBN 1-879478-25-0

Contents

Why Memorize Math Facts? v

Getting Started: Tips for Teachers Using this Book vi

Timed Test Masters 1

1-2 7

3 9

1-3 11

4 13

1-4 15

5 17

1-5 19

6 21

1-6 23

7 25

1-7 27

8 29

1-8 31

9 33

1-9 35

10 37

1-10 39

Folders and Records 49

Division Sticker Sheet 52

Record of Timed Tests 53

Graph 54

Teacher Record 55

Congratulations Certificates 56

Flash Card Masters 57

Review Masters 79

Addition 81

Subtraction 83

Multiplication 85

Why Memorize Math Facts?

There are many benefits to memorizing basic math facts. With the methods and materials in this book and its companion books, an investment of five minutes a day in grades 2 (addition), 3 (subtraction), 4 (multiplication) and 5 (division), will pay dividends for a lifetime, *conserving* time and effort in the long run. Here are some benefits of memorizing math facts:

- Knowing basic facts frees students' minds to concentrate fully on math concepts, skills, and problem-solving. Overlearned, mastered facts become background information, immediately available but requiring little or no conscious thought.
- Math is exact. A wrong answer because of a missed fact is no less wrong than a wrong answer because a concept isn't understood.
- After students have mastered the basic facts, teachers can progress much more quickly through the curriculum. Working with students who *don't* know their facts at higher grade levels places a drag on the process of teaching mathematical concepts and skills.
- Because the facts are so basic and so useful in math, students who know them have a much better foundation for junior high, high school, and even college than those who do not.
- Although calculators and computers are wonderful tools, a person who must locate a calculator in order to figure out $10 \div 5$ would not be considered mathematically literate by most, regardless of his/her understanding of concepts.
- Knowing the basic facts gives students a sense of self-esteem. Students who know the facts generally feel more confident of themselves and have a more positive attitude toward math than students who do not.
- Memorizing facts need not interfere in any way with teaching concepts, problem-solving, etc. It is a separate activity which requires five minutes a day in grades 2-5, leaving the rest of math time for teaching everything else.

Getting Started
Tips for Teachers Using This Book

- First read p. 2 for a quick explanation of how the five-minutes-a-day activity works and a step-by-step checklist for advance preparation.

- Scan pp. 3-6 for answers to any questions that may have arisen as you read p. 2.

- If you open the book to any page and have questions about what you see, page backwards to the section divider and read the pages immediately following it.

- Worksheets are on the front side of each page. Corresponding answer keys are on the back.

- Teachers are welcomed and encouraged to adapt these materials according to their own preferences and the needs of their students. For example, you may choose to have students pass the 10's worksheet before memorizing the 4's.

TIMED
TEST
MASTERS

TIMED TESTS

The black-line masters in this book are intended to be flexible so that teachers can use them in any way that is appropriate for their students. In this section are ideas and suggestions based on how I use these materials. Adapt as necessary.

Basic Procedure for Timed Tests

1) Students get out their math folders.
2) Worksheets are distributed.
3) I say, "Ready, set, GO!" and start a stopwatch.
4) Students work through the test as quickly and accurately as they can.
5) As each student finishes, a note is made of his/her time.
6) After five minutes, I say, "STOP!"
7) I collect papers from students who have finished in three minutes or less and check them myself. Anyone with 100 facts correct in three minutes or less "passes."
8) Other students check their own tests and record their scores and times in their folders.

Step-by-Step Checklist for Preparing to Use this Basic Procedure

❐ Read the introductory information on pp. 2-6. Highlight key points for easy reference later!

❐ Label a file folder for each of the timed tests on pp. 7-47 and 81-87.

❐ Reproduce numerous copies of each of the timed tests (not keys) and place them in the appropriate folders. For a classroom, at least 100 copies of each test will likely be needed. To conserve paper, copy on the back of waste paper or make two-sided copies, then recycle.

❐ Prepare a folder for each student. Instructions and reproducible masters are on pp. 50-54. Staple a sticker sheet on the left (p. 52) and either the Record of Timed tests or a graph on the right (pp. 53-54; instructions for preparation before duplicating are on p. 50).

❐ Reproduce answer keys for each student on colored paper. Answer keys are on the backs of each test master, pp. 8-48 and 82-88. Staple packets of keys together and place in folders.

❐ Prepare or purchase a chart of multiples to post in back of the classroom, **or** reproduce the chart of facts on p. 51 and tape it in each folder.

❐ Read the fourth paragraph on p. 4 and decide how you want to keep times. Get a stop watch **or** prepare a time chart as shown on p. 51 and get a pointer.

❐ Buy stickers to award after individual tests are passed and colorful certificates to award to students who have memorized *all* addition, subtraction, or multiplication facts to mastery. Decide what other special reward you might offer when mastery of *all* facts, including division facts, is accomplished.

❐ Reproduce and cut apart numerous congratulations certificates (p. 56). Store them in an accessible place. These go home after individual tests are passed, decorated with stickers.

❐ Optional: Reproduce personal flash cards for each student's use as sets of facts are introduced (pp. 59-78). These can be stored in plastic sandwich bags in the folders. They will not be needed until review of addition, subtraction, and multiplication is complete.

❐ Optional: Reproduce a special set of review tests for multiplication facts (p. 85 or 87) for a substitute to use if you are unexpectedly absent.

Reviewing Addition, Subtraction, and Multiplication Facts

Pretest
- Explain the importance of memorizing basic math facts. See p. v for reasons it's important.
- Distribute the addition timed test (p. 81) to all students.
- Time students for three minutes.
- Collect and check the papers. This shows the teacher each individual's starting point.

Folders
- Distribute folders and have students write their names on them.
- Again, distribute the addition timed test (p. 81) to all students.
- Explain how times will be kept. If you use a time chart, instruct students to look at the chart as soon as they finish and write the time you are pointing to at that time (see p. 51). If you use a stopwatch, students must raise their hands as soon as they finish and you note the time. In this case, you will read times back to students as soon as the timed test is over.
- Time students for three minutes.
- Collect folders from students who finish in three minutes or less and check their tests as soon as time allows. If 100 facts are correct, award a colorful certificate and applaud! They can continue doing the addition worksheet for better times or go on to subtraction. See p. 6, paragraph 4.
- Read the answers aloud as everyone else checks their own test **or** distribute copies of answer keys for students to use while checking.
- Students graph or record their scores in the appropriate place in their folder.

Continue review tests until all or nearly all students have mastered addition, subtraction and multiplication facts.

Division Timed Tests

Introducing a Set of Facts
- Demonstrate or have students generate a set of facts using manipulatives.
- Distribute personal flash cards and sandwich bags for the sets introduced (pp. 59-78).
- Have students practice with flash cards alone or with a partner for several days.

Division Timed Tests
- Timed tests are the first activity in my math class. Students have folders out and ready.
- Distribute copies of the answer keys for division if students don't already have them.
- Distribute copies of the timed tests. The first day everyone will have ones and twos (p. 7).
- Time students for five minutes.
- Collect folders and tests from everyone who finished in three minutes or less.
- Other students check their own papers and record their times and scores in their folders.

- Many students pass the first day in three minutes or less. I heartily congratulate them, and emphasize that they've already mastered 20 of the 100 basic division facts. There are only 80 to go and this is just the first *day*. We have until the end of the *year* to do it. Although other facts will be harder than ones and twos to memorize, this is an excellent start! This **early success** gives students confidence. A certificate goes home, a sticker is awarded, and the wall chart is marked.

- For those who don't pass the first day: Don't worry. You'll have another chance tomorrow and the next day and the next. This is not a race. You have a whole *year* to learn these facts. It is naturally easier for some kids than for others to memorize these because some people are fortunate to be born with mathematical talent. If math isn't your area of talent, you have *other* areas. Learning the facts might take you longer than someone else, but it doesn't matter how long it takes, it just matters that you *learn* them by the end of the year. If you do, you'll be *way ahead* of many kids in sixth grade around the United States. There are a lot of *junior high and high school* students who don't know their facts well. It might be harder for you than for someone else, but *you can do it!* Persistence will pay off and the more difficult it is for you, the more credit you'll deserve when you accomplish it!

- Most test papers will be used and then discarded, so conservation may be a concern. I ask parents to be on the lookout for paper which has been printed on one side and is to be discarded. In this way, I "precycle" paper (which we then "recycle" as a class). I rarely need to use fresh paper, and when I do, I print tests on both sides.

- All students who have passed ones and twos do the threes worksheet the next day. If they pass it in three minutes or less with 100% accuracy, they do the 1-3's worksheet the next day. Eventually, each student will be working at a level which challenges him/her. To distribute the worksheets, I walk around the room with a stack of file folders containing blank worksheets and ask, "Who needs 3's? Who needs 1-3's?" and so on. This takes only a few minutes each day.

- Recording the times can be done in any of several ways. I sit at my desk with a stopwatch and scan the room. As soon as a student finishes, he raises his hand and I record his initials and the time. I read these aloud as soon as the five minutes are up and students note their own times in their folders. Another method is to make a large time chart as shown on p. 51. The teacher stands next to the chart and points to each time as it elapses on the stopwatch or a wall clock. When a student is finished, she looks up and records the time to which the teacher is pointing. Another possibility is to have students use their own stopwatches.

- I have students do their own checking and record-keeping until they complete 100 facts in three minutes or less. When they do, I collect the entire folder including the test. There are only a few of these each day. I check these tests, and if all 100 facts are correct, I record the date on a chart such as that on p. 55, give a sticker (p. 52), and prepare a certificate to go home (p. 56). If any facts were missed, the test must be tried again.

- When a student passes a final worksheet (Division I, II, III, IV or V on pp. 39-47), there is a BIG reward. I give a nice certificate with a Susan B. Anthony dollar. This costs me $20+ dollars a year, but it is *well worth the cost!* Some students spend their dollars immediately. Others keep them as souvenirs. On a schoolwide basis, there can be recognition at an awards assembly or a mention in the school newspaper. Emphasize to students who haven't yet passed that their time will come! Their reward is ready and waiting for them and you have confidence that they'll earn it!

Questions and Answers

- **What about students who can't handle time pressure or have test anxiety?** Occasionally, early in the year, a student will panic and "freeze" during a timed test. If this happens, I counsel with the student. I tell her that even though we call this a "test" it is not graded and there is no way to fail. If she got only ten answers correct today and can get eleven correct tomorrow, that is progress and will eventually lead to her knowing all the facts. There is no hurry. We have all year. As long as she's headed in the right direction, she can relax and just do her best. Some parents have told me that their children initially felt threatened by timed tests but got over it after a short time and began to like them. In the process, they lost some of their fear of tests in other subject areas.

- **What about the aspect of competition?** I post a wall chart such as that shown on p. 55. Children know where they stand in relation to others. I *continually* emphasize that learning math facts is not a race, that it is easier for some because they were lucky enough to be born with talent in this area, and that the goal is for *everyone* to do 100 facts in three minutes or less by the end of the *year*. In my experience, the top students do race and compete with each other. That is one reason they like the tests and aren't bored by them. I encourage students to support and cheer for each other, and they do. When a student who's had difficulty passes a test, she receives more cheers and applause than the top students! When parents come to the classroom, kids at all levels often escort them directly to the wall chart to show how they're doing. This program will work with or without a wall chart, but in my experience, this can be handled so that students benefit from whatever competition naturally takes place.

- **Does this activity ever get boring?** It is a routine, rather like brushing one's teeth or having lunch at a particular time. In my experience, students do not become bored, possibly because as soon as they master one worksheet, they move on to another. Those who have passed *all* worksheets are given other options. In my experience, many of them choose to continue doing the timed tests even when they are free to choose to do something else.

- **What about special education students or others who just can't succeed at this?** I have had a few students who were truly unable to complete a worksheet in three minutes. Once they got behind and were not making progress, I allowed them to pass on to the next worksheet by getting 100 correct in *five* minutes or less. To get the final reward for knowing *all* the facts (a Susan B. Anthony dollar in my case), they had to pass in three minutes or less like everyone else. There were students who had a great deal of trouble early in the year, who went on the five minute plan and passed all of the worksheets, and who then worked on speed and passed the final test in three minutes for their reward. I escorted them straight to the principal with their papers. Parents were called immediately with the good news. I emphasized to the whole class that this had not been easy, that this child had persisted and succeeded despite the frustration he'd felt, that *anyone* can persist and succeed at things that are difficult for them. I emphasize to the child that he is now far *ahead* of many older students who don't know their facts. A *very* few students did not achieve the goal of 100 facts correct in three minutes or less during the year. However, *these students still knew the facts!* I had achieved *my* objective. Even students who find math difficult like the timed tests and are proud of the measurable progress they make.

- **What about students who can't write fast?** Students who are slow writers may pass the tests one-on-one orally with me when they are ready. Generally, this is after I talk with parents.

- **What about left-handed students?** Left-handed students may be frustrated if their hand covers the question. Allow them to write the answers in the column to the *left* of the question.

- **Does being "behind" harm a student's self-esteem?** Although teachers may avoid comparing students, children already know that they are better in some subject areas than others. They already know who is doing better and worse in various subjects than they are. Achieving an objective goal such as getting 100 facts correct in three minutes, despite the frustration and discouragement that may occur along the way, is an excellent way to *enhance* self-esteem. Children learn that they can achieve things they may have initially thought impossible.

- **What about the students who already know the facts?** During review of addition, subtraction, and multiplication facts, some children may pass in less than three minutes the first day. They can then move on or work to improve their speeds, with the idea of challenging the principal or older students to a speed test one day. This keeps the tests interesting for top students. In division, students may proceed through the worksheets as quickly as one per day (finishing in as little as three weeks). They then can choose to improve their speed on the final tests. Students who have passed all tests have the option of doing only one division test per week, to maintain mastery, and using the time on the other four days to read or do special activities. They often prefer to continue doing the tests. This is an area in which they shine and they like it!

- **What do students do when they finish early and are waiting for the five minutes to elapse?** They can go over their papers to make sure they didn't make any careless errors. They can use the key to check and score their paper. They can read or do a quiet activity. They can quietly encourage a nearby friend.

- **What about cheating?** In my experience, there has been little or no cheating. The teacher checks any papers which count for "passing" so there is little incentive to cheat. If cheating does occur, I have a talk with the student to express my disappointment and emphasize that the goal is to learn the facts, not pass the tests. I offer to do whatever I can to help the child learn the facts. If the problem were to recur (it never has), parents would be called.

- **How can parents help?** I send home flashcards after we've used them in class, and at the first conference I provide parents with a photocopied set of all the timed tests, which they can copy and use at home. Some children like to do a "practice test" every night at home!

- **What about days when a substitute is in the room?** Timed tests are difficult for a substitute to manage once students are progressing individually. If I know I will be gone, I either pass out the appropriate worksheets the day before or write student names on the needed worksheets so a substitute can just keep time and collect the folders. Another option is to duplicate a set of review multiplication tests and offer something special to anyone who remembers all of them well enough to pass in three minutes or less. My students were quite disappointed when they didn't get to do the tests, so warn students in advance not to expect regular timed tests when a substitute is teaching.

Division 1-2

Name_____ Date _____ Score _____ Time _____

(1) $14 \div 2$		(26) $6 \div 2$		(51) $12 \div 2$		(76) $4 \div 1$	
(2) $9 \div 1$		(27) $4 \div 1$		(52) $16 \div 2$		(77) $7 \div 1$	
(3) $2 \div 2$		(28) $4 \div 2$		(53) $5 \div 1$		(78) $14 \div 2$	
(4) $12 \div 2$		(29) $9 \div 1$		(54) $14 \div 2$		(79) $18 \div 2$	
(5) $10 \div 1$		(30) $12 \div 2$		(55) $8 \div 1$		(80) $6 \div 2$	
(6) $5 \div 1$		(31) $3 \div 1$		(56) $10 \div 2$		(81) $9 \div 1$	
(7) $4 \div 2$		(32) $20 \div 2$		(57) $4 \div 1$		(82) $1 \div 1$	
(8) $2 \div 1$		(33) $6 \div 1$		(58) $16 \div 2$		(83) $2 \div 1$	
(9) $18 \div 2$		(34) $10 \div 2$		(59) $1 \div 1$		(84) $12 \div 2$	
(10) $8 \div 1$		(35) $5 \div 1$		(60) $8 \div 2$		(85) $6 \div 1$	
(11) $6 \div 2$		(36) $12 \div 2$		(61) $3 \div 1$		(86) $2 \div 2$	
(12) $4 \div 1$		(37) $18 \div 2$		(62) $6 \div 2$		(87) $8 \div 1$	
(13) $12 \div 2$		(38) $7 \div 1$		(63) $10 \div 1$		(88) $14 \div 2$	
(14) $6 \div 1$		(39) $10 \div 2$		(64) $4 \div 2$		(89) $20 \div 2$	
(15) $10 \div 1$		(40) $1 \div 1$		(65) $9 \div 1$		(90) $4 \div 2$	
(16) $3 \div 1$		(41) $14 \div 2$		(66) $20 \div 2$		(91) $16 \div 2$	
(17) $16 \div 2$		(42) $8 \div 1$		(67) $10 \div 1$		(92) $10 \div 1$	
(18) $5 \div 1$		(43) $2 \div 2$		(68) $6 \div 1$		(93) $8 \div 2$	
(19) $8 \div 2$		(44) $16 \div 2$		(69) $2 \div 2$		(94) $3 \div 1$	
(20) $7 \div 1$		(45) $2 \div 1$		(70) $12 \div 2$		(95) $5 \div 1$	
(21) $10 \div 2$		(46) $8 \div 2$		(71) $9 \div 1$		(96) $12 \div 2$	
(22) $1 \div 1$		(47) $10 \div 1$		(72) $8 \div 2$		(97) $8 \div 2$	
(23) $20 \div 2$		(48) $5 \div 1$		(73) $2 \div 1$		(98) $8 \div 1$	
(24) $2 \div 1$		(49) $18 \div 2$		(74) $7 \div 1$		(99) $4 \div 2$	
(25) $6 \div 1$		(50) $6 \div 2$		(75) $18 \div 2$		(100) $16 \div 2$	

Division 1-2

(1) $14 \div 2$	7	(26) $6 \div 2$	3	(51) $12 \div 2$	6	(76) $4 \div 1$	4
(2) $9 \div 1$	9	(27) $4 \div 1$	4	(52) $16 \div 2$	8	(77) $7 \div 1$	7
(3) $2 \div 2$	1	(28) $4 \div 2$	2	(53) $5 \div 1$	5	(78) $14 \div 2$	7
(4) $12 \div 2$	6	(29) $9 \div 1$	9	(54) $14 \div 2$	7	(79) $18 \div 2$	9
(5) $10 \div 1$	10	(30) $12 \div 2$	6	(55) $8 \div 1$	8	(80) $6 \div 2$	3
(6) $5 \div 1$	5	(31) $3 \div 1$	3	(56) $10 \div 2$	5	(81) $9 \div 1$	9
(7) $4 \div 2$	2	(32) $20 \div 2$	10	(57) $4 \div 1$	4	(82) $1 \div 1$	1
(8) $2 \div 1$	2	(33) $6 \div 1$	6	(58) $16 \div 2$	8	(83) $2 \div 1$	2
(9) $18 \div 2$	9	(34) $10 \div 2$	5	(59) $1 \div 1$	1	(84) $12 \div 2$	6
(10) $8 \div 1$	8	(35) $5 \div 1$	5	(60) $8 \div 2$	4	(85) $6 \div 1$	6
(11) $6 \div 2$	3	(36) $12 \div 2$	6	(61) $3 \div 1$	3	(86) $2 \div 2$	1
(12) $4 \div 1$	4	(37) $18 \div 2$	9	(62) $6 \div 2$	3	(87) $8 \div 1$	8
(13) $12 \div 2$	6	(38) $7 \div 1$	7	(63) $10 \div 1$	10	(88) $14 \div 2$	7
(14) $6 \div 1$	6	(39) $10 \div 2$	5	(64) $4 \div 2$	2	(89) $20 \div 2$	10
(15) $10 \div 1$	10	(40) $1 \div 1$	1	(65) $9 \div 1$	9	(90) $4 \div 2$	2
(16) $3 \div 1$	3	(41) $14 \div 2$	7	(66) $20 \div 2$	10	(91) $16 \div 2$	8
(17) $16 \div 2$	8	(42) $8 \div 1$	8	(67) $10 \div 1$	10	(92) $10 \div 1$	10
(18) $5 \div 1$	5	(43) $2 \div 2$	1	(68) $6 \div 1$	6	(93) $8 \div 2$	4
(19) $8 \div 2$	4	(44) $16 \div 2$	8	(69) $2 \div 2$	1	(94) $3 \div 1$	3
(20) $7 \div 1$	7	(45) $2 \div 1$	2	(70) $12 \div 2$	6	(95) $5 \div 1$	5
(21) $10 \div 2$	5	(46) $8 \div 2$	4	(71) $9 \div 1$	9	(96) $12 \div 2$	6
(22) $1 \div 1$	1	(47) $10 \div 1$	10	(72) $8 \div 2$	4	(97) $8 \div 2$	4
(23) $20 \div 2$	10	(48) $5 \div 1$	5	(73) $2 \div 1$	2	(98) $8 \div 1$	8
(24) $2 \div 1$	2	(49) $18 \div 2$	9	(74) $7 \div 1$	7	(99) $4 \div 2$	2
(25) $6 \div 1$	6	(50) $6 \div 2$	3	(75) $18 \div 2$	9	(100) $16 \div 2$	8

Division 3

Name_____ Date_____ Score_____ Time_____

(1) $9 \div 3$	(26) $12 \div 3$	(51) $18 \div 3$	(76) $27 \div 3$
(2) $21 \div 3$	(27) $6 \div 3$	(52) $9 \div 3$	(77) $30 \div 3$
(3) $30 \div 3$	(28) $9 \div 3$	(53) $3 \div 3$	(78) $3 \div 3$
(4) $18 \div 3$	(29) $30 \div 3$	(54) $6 \div 3$	(79) $18 \div 3$
(5) $27 \div 3$	(30) $18 \div 3$	(55) $21 \div 3$	(80) $9 \div 3$
(6) $12 \div 3$	(31) $15 \div 3$	(56) $15 \div 3$	(81) $6 \div 3$
(7) $3 \div 3$	(32) $27 \div 3$	(57) $12 \div 3$	(82) $15 \div 3$
(8) $24 \div 3$	(33) $3 \div 3$	(58) $27 \div 3$	(83) $24 \div 3$
(9) $6 \div 3$	(34) $15 \div 3$	(59) $30 \div 3$	(84) $12 \div 3$
(10) $15 \div 3$	(35) $24 \div 3$	(60) $3 \div 3$	(85) $21 \div 3$
(11) $21 \div 3$	(36) $21 \div 3$	(61) $24 \div 3$	(86) $18 \div 3$
(12) $18 \div 3$	(37) $6 \div 3$	(62) $6 \div 3$	(87) $27 \div 3$
(13) $3 \div 3$	(38) $30 \div 3$	(63) $12 \div 3$	(88) $12 \div 3$
(14) $27 \div 3$	(39) $24 \div 3$	(64) $18 \div 3$	(89) $30 \div 3$
(15) $12 \div 3$	(40) $15 \div 3$	(65) $9 \div 3$	(90) $9 \div 3$
(16) $30 \div 3$	(41) $27 \div 3$	(66) $24 \div 3$	(91) $18 \div 3$
(17) $9 \div 3$	(42) $12 \div 3$	(67) $30 \div 3$	(92) $3 \div 3$
(18) $18 \div 3$	(43) $6 \div 3$	(68) $21 \div 3$	(93) $21 \div 3$
(19) $15 \div 3$	(44) $18 \div 3$	(69) $15 \div 3$	(94) $12 \div 3$
(20) $3 \div 3$	(45) $24 \div 3$	(70) $30 \div 3$	(95) $15 \div 3$
(21) $21 \div 3$	(46) $3 \div 3$	(71) $27 \div 3$	(96) $9 \div 3$
(22) $6 \div 3$	(47) $21 \div 3$	(72) $9 \div 3$	(97) $24 \div 3$
(23) $24 \div 3$	(48) $9 \div 3$	(73) $3 \div 3$	(98) $6 \div 3$
(24) $15 \div 3$	(49) $21 \div 3$	(74) $12 \div 3$	(99) $27 \div 3$
(25) $6 \div 3$	(50) $30 \div 3$	(75) $24 \div 3$	(100) $3 \div 3$

Division 3

(1) 9 ÷ 3	3	(26) 12 ÷ 3	4	(51) 18 ÷ 3	6	(76) 27 ÷ 3	9
(2) 21 ÷ 3	7	(27) 6 ÷ 3	2	(52) 9 ÷ 3	3	(77) 30 ÷ 3	10
(3) 30 ÷ 3	10	(28) 9 ÷ 3	3	(53) 3 ÷ 3	1	(78) 3 ÷ 3	1
(4) 18 ÷ 3	6	(29) 30 ÷ 3	10	(54) 6 ÷ 3	2	(79) 18 ÷ 3	6
(5) 27 ÷ 3	9	(30) 18 ÷ 3	6	(55) 21 ÷ 3	7	(80) 9 ÷ 3	3
(6) 12 ÷ 3	4	(31) 15 ÷ 3	5	(56) 15 ÷ 3	5	(81) 6 ÷ 3	2
(7) 3 ÷ 3	1	(32) 27 ÷ 3	9	(57) 12 ÷ 3	4	(82) 15 ÷ 3	5
(8) 24 ÷ 3	8	(33) 3 ÷ 3	1	(58) 27 ÷ 3	9	(83) 24 ÷ 3	8
(9) 6 ÷ 3	2	(34) 15 ÷ 3	5	(59) 30 ÷ 3	10	(84) 12 ÷ 3	4
(10) 15 ÷ 3	5	(35) 24 ÷ 3	8	(60) 3 ÷ 3	1	(85) 21 ÷ 3	7
(11) 21 ÷ 3	7	(36) 21 ÷ 3	7	(61) 24 ÷ 3	8	(86) 18 ÷ 3	6
(12) 18 ÷ 3	6	(37) 6 ÷ 3	2	(62) 6 ÷ 3	2	(87) 27 ÷ 3	9
(13) 3 ÷ 3	1	(38) 30 ÷ 3	10	(63) 12 ÷ 3	4	(88) 12 ÷ 3	4
(14) 27 ÷ 3	9	(39) 24 ÷ 3	8	(64) 18 ÷ 3	6	(89) 30 ÷ 3	10
(15) 12 ÷ 3	4	(40) 15 ÷ 3	5	(65) 9 ÷ 3	3	(90) 9 ÷ 3	3
(16) 30 ÷ 3	10	(41) 27 ÷ 3	9	(66) 24 ÷ 3	8	(91) 18 ÷ 3	6
(17) 9 ÷ 3	3	(42) 12 ÷ 3	4	(67) 30 ÷ 3	10	(92) 3 ÷ 3	1
(18) 18 ÷ 3	6	(43) 6 ÷ 3	2	(68) 21 ÷ 3	7	(93) 21 ÷ 3	7
(19) 15 ÷ 3	5	(44) 18 ÷ 3	6	(69) 15 ÷ 3	5	(94) 12 ÷ 3	4
(20) 3 ÷ 3	1	(45) 24 ÷ 3	8	(70) 30 ÷ 3	10	(95) 15 ÷ 3	5
(21) 21 ÷ 3	7	(46) 3 ÷ 3	1	(71) 27 ÷ 3	9	(96) 9 ÷ 3	3
(22) 6 ÷ 3	2	(47) 21 ÷ 3	7	(72) 9 ÷ 3	3	(97) 24 ÷ 3	8
(23) 24 ÷ 3	8	(48) 9 ÷ 3	3	(73) 3 ÷ 3	1	(98) 6 ÷ 3	2
(24) 15 ÷ 3	5	(49) 21 ÷ 3	7	(74) 12 ÷ 3	4	(99) 27 ÷ 3	9
(25) 6 ÷ 3	2	(50) 30 ÷ 3	10	(75) 24 ÷ 3	8	(100) 3 ÷ 3	1

Division 1-3

Name_____ Date _____ Score _____ Time _____

(1) $6 \div 3$		(26) $3 \div 1$		(51) $7 \div 1$		(76) $4 \div 2$	
(2) $18 \div 2$		(27) $10 \div 2$		(52) $30 \div 3$		(77) $3 \div 1$	
(3) $7 \div 1$		(28) $24 \div 3$		(53) $5 \div 1$		(78) $6 \div 3$	
(4) $15 \div 3$		(29) $4 \div 2$		(54) $9 \div 3$		(79) $12 \div 2$	
(5) $12 \div 2$		(30) $1 \div 1$		(55) $20 \div 2$		(80) $21 \div 3$	
(6) $30 \div 3$		(31) $18 \div 2$		(56) $2 \div 2$		(81) $10 \div 1$	
(7) $2 \div 1$		(32) $12 \div 2$		(57) $8 \div 2$		(82) $27 \div 3$	
(8) $21 \div 3$		(33) $21 \div 3$		(58) $1 \div 1$		(83) $20 \div 2$	
(9) $5 \div 1$		(34) $10 \div 1$		(59) $3 \div 1$		(84) $12 \div 3$	
(10) $6 \div 2$		(35) $4 \div 1$		(60) $14 \div 2$		(85) $16 \div 2$	
(11) $10 \div 1$		(36) $12 \div 3$		(61) $9 \div 1$		(86) $3 \div 3$	
(12) $9 \div 3$		(37) $9 \div 1$		(62) $2 \div 2$		(87) $8 \div 2$	
(13) $27 \div 3$		(38) $6 \div 1$		(63) $8 \div 1$		(88) $18 \div 3$	
(14) $4 \div 1$		(39) $18 \div 3$		(64) $27 \div 3$		(89) $10 \div 2$	
(15) $20 \div 2$		(40) $14 \div 2$		(65) $10 \div 1$		(90) $24 \div 3$	
(16) $8 \div 1$		(41) $16 \div 2$		(66) $21 \div 3$		(91) $1 \div 1$	
(17) $12 \div 3$		(42) $8 \div 1$		(67) $30 \div 3$		(92) $12 \div 2$	
(18) $2 \div 2$		(43) $27 \div 3$		(68) $15 \div 3$		(93) $21 \div 3$	
(19) $16 \div 2$		(44) $6 \div 2$		(69) $18 \div 2$		(94) $2 \div 1$	
(20) $9 \div 1$		(45) $2 \div 1$		(70) $6 \div 3$		(95) $6 \div 1$	
(21) $3 \div 3$		(46) $15 \div 3$		(71) $7 \div 1$		(96) $9 \div 1$	
(22) $14 \div 2$		(47) $6 \div 3$		(72) $12 \div 2$		(97) $10 \div 2$	
(23) $6 \div 1$		(48) $10 \div 2$		(73) $2 \div 1$		(98) $15 \div 3$	
(24) $8 \div 2$		(49) $4 \div 2$		(74) $6 \div 2$		(99) $4 \div 2$	
(25) $18 \div 3$		(50) $24 \div 3$		(75) $5 \div 1$		(100) $8 \div 1$	

Division 1-3

#	Problem	Answer
(1)	6 ÷ 3	2
(2)	18 ÷ 2	9
(3)	7 ÷ 1	7
(4)	15 ÷ 3	5
(5)	12 ÷ 2	6
(6)	30 ÷ 3	10
(7)	2 ÷ 1	2
(8)	21 ÷ 3	7
(9)	5 ÷ 1	5
(10)	6 ÷ 2	3
(11)	10 ÷ 1	10
(12)	9 ÷ 3	3
(13)	27 ÷ 3	9
(14)	4 ÷ 1	4
(15)	20 ÷ 2	10
(16)	8 ÷ 1	8
(17)	12 ÷ 3	4
(18)	2 ÷ 2	1
(19)	16 ÷ 2	8
(20)	9 ÷ 1	9
(21)	3 ÷ 3	1
(22)	14 ÷ 2	7
(23)	6 ÷ 1	6
(24)	8 ÷ 2	4
(25)	18 ÷ 3	6
(26)	3 ÷ 1	3
(27)	10 ÷ 2	5
(28)	24 ÷ 3	8
(29)	4 ÷ 2	2
(30)	1 ÷ 1	1
(31)	18 ÷ 2	9
(32)	12 ÷ 2	6
(33)	21 ÷ 3	7
(34)	10 ÷ 1	10
(35)	4 ÷ 1	4
(36)	12 ÷ 3	4
(37)	9 ÷ 1	9
(38)	6 ÷ 1	6
(39)	18 ÷ 3	6
(40)	14 ÷ 2	7
(41)	16 ÷ 2	8
(42)	8 ÷ 1	8
(43)	27 ÷ 3	9
(44)	6 ÷ 2	3
(45)	2 ÷ 1	2
(46)	15 ÷ 3	5
(47)	6 ÷ 3	2
(48)	10 ÷ 2	5
(49)	4 ÷ 2	2
(50)	24 ÷ 3	8
(51)	7 ÷ 1	7
(52)	30 ÷ 3	10
(53)	5 ÷ 1	5
(54)	9 ÷ 3	3
(55)	20 ÷ 2	10
(56)	2 ÷ 2	1
(57)	8 ÷ 2	4
(58)	1 ÷ 1	1
(59)	3 ÷ 1	3
(60)	14 ÷ 2	7
(61)	9 ÷ 1	9
(62)	2 ÷ 2	1
(63)	8 ÷ 1	8
(64)	27 ÷ 3	9
(65)	10 ÷ 1	10
(66)	21 ÷ 3	7
(67)	30 ÷ 3	10
(68)	15 ÷ 3	5
(69)	18 ÷ 2	9
(70)	6 ÷ 3	2
(71)	7 ÷ 1	7
(72)	12 ÷ 2	6
(73)	2 ÷ 1	2
(74)	6 ÷ 2	3
(75)	5 ÷ 1	5
(76)	4 ÷ 2	2
(77)	3 ÷ 1	3
(78)	6 ÷ 3	2
(79)	12 ÷ 2	6
(80)	21 ÷ 3	7
(81)	10 ÷ 1	10
(82)	27 ÷ 3	9
(83)	20 ÷ 2	10
(84)	12 ÷ 3	4
(85)	16 ÷ 2	8
(86)	3 ÷ 3	1
(87)	8 ÷ 2	4
(88)	18 ÷ 3	6
(89)	10 ÷ 2	5
(90)	24 ÷ 3	8
(91)	1 ÷ 1	1
(92)	12 ÷ 2	6
(93)	21 ÷ 3	7
(94)	2 ÷ 1	2
(95)	6 ÷ 1	6
(96)	9 ÷ 1	9
(97)	10 ÷ 2	5
(98)	15 ÷ 3	5
(99)	4 ÷ 2	2
(100)	8 ÷ 1	8

Division 4

Name_____ Date _____ Score _____ Time _____

(1) $36 \div 4$		(26) $24 \div 4$		(51) $16 \div 4$		(76) $40 \div 4$	
(2) $16 \div 4$		(27) $20 \div 4$		(52) $32 \div 4$		(77) $8 \div 4$	
(3) $40 \div 4$		(28) $36 \div 4$		(53) $20 \div 4$		(78) $4 \div 4$	
(4) $8 \div 4$		(29) $12 \div 4$		(54) $24 \div 4$		(79) $16 \div 4$	
(5) $24 \div 4$		(30) $28 \div 4$		(55) $40 \div 4$		(80) $20 \div 4$	
(6) $12 \div 4$		(31) $4 \div 4$		(56) $4 \div 4$		(81) $36 \div 4$	
(7) $32 \div 4$		(32) $16 \div 4$		(57) $28 \div 4$		(82) $24 \div 4$	
(8) $20 \div 4$		(33) $40 \div 4$		(58) $12 \div 4$		(83) $28 \div 4$	
(9) $28 \div 4$		(34) $8 \div 4$		(59) $36 \div 4$		(84) $4 \div 4$	
(10) $4 \div 4$		(35) $32 \div 4$		(60) $20 \div 4$		(85) $32 \div 4$	
(11) $36 \div 4$		(36) $28 \div 4$		(61) $8 \div 4$		(86) $16 \div 4$	
(12) $16 \div 4$		(37) $24 \div 4$		(62) $40 \div 4$		(87) $12 \div 4$	
(13) $32 \div 4$		(38) $12 \div 4$		(63) $4 \div 4$		(88) $4 \div 4$	
(14) $28 \div 4$		(39) $36 \div 4$		(64) $24 \div 4$		(89) $24 \div 4$	
(15) $40 \div 4$		(40) $20 \div 4$		(65) $8 \div 4$		(90) $8 \div 4$	
(16) $12 \div 4$		(41) $4 \div 4$		(66) $32 \div 4$		(91) $40 \div 4$	
(17) $8 \div 4$		(42) $28 \div 4$		(67) $16 \div 4$		(92) $32 \div 4$	
(18) $36 \div 4$		(43) $12 \div 4$		(68) $20 \div 4$		(93) $12 \div 4$	
(19) $24 \div 4$		(44) $32 \div 4$		(69) $36 \div 4$		(94) $24 \div 4$	
(20) $4 \div 4$		(45) $16 \div 4$		(70) $8 \div 4$		(95) $28 \div 4$	
(21) $32 \div 4$		(46) $24 \div 4$		(71) $12 \div 4$		(96) $20 \div 4$	
(22) $20 \div 4$		(47) $40 \div 4$		(72) $28 \div 4$		(97) $36 \div 4$	
(23) $28 \div 4$		(48) $8 \div 4$		(73) $4 \div 4$		(98) $16 \div 4$	
(24) $40 \div 4$		(49) $20 \div 4$		(74) $8 \div 4$		(99) $32 \div 4$	
(25) $16 \div 4$		(50) $28 \div 4$		(75) $36 \div 4$		(100) $40 \div 4$	

14

Division 4

(1) $36 \div 4$	9	(26) $24 \div 4$	6	(51) $16 \div 4$	4	(76) $40 \div 4$	10
(2) $16 \div 4$	4	(27) $20 \div 4$	5	(52) $32 \div 4$	8	(77) $8 \div 4$	2
(3) $40 \div 4$	10	(28) $36 \div 4$	9	(53) $20 \div 4$	5	(78) $4 \div 4$	1
(4) $8 \div 4$	2	(29) $12 \div 4$	3	(54) $24 \div 4$	6	(79) $16 \div 4$	4
(5) $24 \div 4$	6	(30) $28 \div 4$	7	(55) $40 \div 4$	10	(80) $20 \div 4$	5
(6) $12 \div 4$	3	(31) $4 \div 4$	1	(56) $4 \div 4$	1	(81) $36 \div 4$	9
(7) $32 \div 4$	8	(32) $16 \div 4$	4	(57) $28 \div 4$	7	(82) $24 \div 4$	6
(8) $20 \div 4$	5	(33) $40 \div 4$	10	(58) $12 \div 4$	3	(83) $28 \div 4$	7
(9) $28 \div 4$	7	(34) $8 \div 4$	2	(59) $36 \div 4$	9	(84) $4 \div 4$	1
(10) $4 \div 4$	1	(35) $32 \div 4$	8	(60) $20 \div 4$	5	(85) $32 \div 4$	8
(11) $36 \div 4$	9	(36) $28 \div 4$	7	(61) $8 \div 4$	2	(86) $16 \div 4$	4
(12) $16 \div 4$	4	(37) $24 \div 4$	6	(62) $40 \div 4$	10	(87) $12 \div 4$	3
(13) $32 \div 4$	8	(38) $12 \div 4$	3	(63) $4 \div 4$	1	(88) $4 \div 4$	1
(14) $28 \div 4$	7	(39) $36 \div 4$	9	(64) $24 \div 4$	6	(89) $24 \div 4$	6
(15) $40 \div 4$	10	(40) $20 \div 4$	5	(65) $8 \div 4$	2	(90) $8 \div 4$	2
(16) $12 \div 4$	3	(41) $4 \div 4$	1	(66) $32 \div 4$	8	(91) $40 \div 4$	10
(17) $8 \div 4$	2	(42) $28 \div 4$	7	(67) $16 \div 4$	4	(92) $32 \div 4$	8
(18) $36 \div 4$	9	(43) $12 \div 4$	3	(68) $20 \div 4$	5	(93) $12 \div 4$	3
(19) $24 \div 4$	6	(44) $32 \div 4$	8	(69) $36 \div 4$	9	(94) $24 \div 4$	6
(20) $4 \div 4$	1	(45) $16 \div 4$	4	(70) $8 \div 4$	2	(95) $28 \div 4$	7
(21) $32 \div 4$	8	(46) $24 \div 4$	6	(71) $12 \div 4$	3	(96) $20 \div 4$	5
(22) $20 \div 4$	5	(47) $40 \div 4$	10	(72) $28 \div 4$	7	(97) $36 \div 4$	9
(23) $28 \div 4$	7	(48) $8 \div 4$	2	(73) $4 \div 4$	1	(98) $16 \div 4$	4
(24) $40 \div 4$	10	(49) $20 \div 4$	5	(74) $8 \div 4$	2	(99) $32 \div 4$	8
(25) $16 \div 4$	4	(50) $28 \div 4$	7	(75) $36 \div 4$	9	(100) $40 \div 4$	10

Division 1-4

Name_____ Date _____ Score _____ Time _____

(1) $12 \div 3$		(26) $32 \div 4$		(51) $40 \div 4$		(76) $3 \div 1$	
(2) $1 \div 1$		(27) $8 \div 2$		(52) $2 \div 1$		(77) $20 \div 2$	
(3) $18 \div 2$		(28) $10 \div 1$		(53) $6 \div 3$		(78) $12 \div 4$	
(4) $8 \div 4$		(29) $3 \div 3$		(54) $14 \div 2$		(79) $20 \div 4$	
(5) $6 \div 1$		(30) $20 \div 4$		(55) $24 \div 3$		(80) $8 \div 2$	
(6) $21 \div 3$		(31) $9 \div 3$		(56) $9 \div 1$		(81) $12 \div 3$	
(7) $24 \div 4$		(32) $27 \div 3$		(57) $2 \div 2$		(82) $24 \div 4$	
(8) $30 \div 3$		(33) $12 \div 4$		(58) $7 \div 1$		(83) $6 \div 2$	
(9) $4 \div 1$		(34) $5 \div 1$		(59) $6 \div 2$		(84) $28 \div 4$	
(10) $8 \div 1$		(35) $18 \div 3$		(60) $36 \div 4$		(85) $2 \div 1$	
(11) $36 \div 4$		(36) $20 \div 2$		(61) $4 \div 1$		(86) $20 \div 4$	
(12) $15 \div 3$		(37) $4 \div 4$		(62) $24 \div 4$		(87) $20 \div 2$	
(13) $6 \div 2$		(38) $12 \div 2$		(63) $6 \div 1$		(88) $18 \div 2$	
(14) $16 \div 2$		(39) $3 \div 1$		(64) $18 \div 2$		(89) $30 \div 3$	
(15) $7 \div 1$		(40) $10 \div 2$		(65) $12 \div 3$		(90) $36 \div 4$	
(16) $16 \div 4$		(41) $1 \div 1$		(66) $10 \div 2$		(91) $16 \div 2$	
(17) $2 \div 2$		(42) $8 \div 4$		(67) $4 \div 4$		(92) $9 \div 1$	
(18) $28 \div 4$		(43) $21 \div 3$		(68) $5 \div 1$		(93) $24 \div 3$	
(19) $9 \div 1$		(44) $30 \div 3$		(69) $9 \div 3$		(94) $14 \div 2$	
(20) $4 \div 2$		(45) $8 \div 1$		(70) $10 \div 1$		(95) $32 \div 4$	
(21) $24 \div 3$		(46) $15 \div 3$		(71) $32 \div 4$		(96) $16 \div 4$	
(22) $40 \div 4$		(47) $16 \div 2$		(72) $3 \div 3$		(97) $15 \div 3$	
(23) $14 \div 2$		(48) $16 \div 4$		(73) $27 \div 3$		(98) $10 \div 1$	
(24) $2 \div 1$		(49) $28 \div 4$		(74) $18 \div 3$		(99) $18 \div 3$	
(25) $6 \div 3$		(50) $4 \div 2$		(75) $12 \div 2$		(100) $6 \div 2$	

Division 1-4

(1)	$12 \div 3$	4	(26)	$32 \div 4$	8	(51)	$40 \div 4$	10	(76)	$3 \div 1$	3
(2)	$1 \div 1$	1	(27)	$8 \div 2$	4	(52)	$2 \div 1$	2	(77)	$20 \div 2$	10
(3)	$18 \div 2$	9	(28)	$10 \div 1$	10	(53)	$6 \div 3$	2	(78)	$12 \div 4$	3
(4)	$8 \div 4$	2	(29)	$3 \div 3$	1	(54)	$14 \div 2$	7	(79)	$20 \div 4$	5
(5)	$6 \div 1$	6	(30)	$20 \div 4$	5	(55)	$24 \div 3$	8	(80)	$8 \div 2$	4
(6)	$21 \div 3$	7	(31)	$9 \div 3$	3	(56)	$9 \div 1$	9	(81)	$12 \div 3$	4
(7)	$24 \div 4$	6	(32)	$27 \div 3$	9	(57)	$2 \div 2$	1	(82)	$24 \div 4$	6
(8)	$30 \div 3$	10	(33)	$12 \div 4$	3	(58)	$7 \div 1$	7	(83)	$6 \div 2$	3
(9)	$4 \div 1$	4	(34)	$5 \div 1$	5	(59)	$6 \div 2$	3	(84)	$28 \div 4$	7
(10)	$8 \div 1$	8	(35)	$18 \div 3$	6	(60)	$36 \div 4$	9	(85)	$2 \div 1$	2
(11)	$36 \div 4$	9	(36)	$20 \div 2$	10	(61)	$4 \div 1$	4	(86)	$20 \div 4$	5
(12)	$15 \div 3$	5	(37)	$4 \div 4$	1	(62)	$24 \div 4$	6	(87)	$20 \div 2$	10
(13)	$6 \div 2$	3	(38)	$12 \div 2$	6	(63)	$6 \div 1$	6	(88)	$18 \div 2$	9
(14)	$16 \div 2$	8	(39)	$3 \div 1$	3	(64)	$18 \div 2$	9	(89)	$30 \div 3$	10
(15)	$7 \div 1$	7	(40)	$10 \div 2$	5	(65)	$12 \div 3$	4	(90)	$36 \div 4$	9
(16)	$16 \div 4$	4	(41)	$1 \div 1$	1	(66)	$10 \div 2$	5	(91)	$16 \div 2$	8
(17)	$2 \div 2$	1	(42)	$8 \div 4$	2	(67)	$4 \div 4$	1	(92)	$9 \div 1$	9
(18)	$28 \div 4$	7	(43)	$21 \div 3$	7	(68)	$5 \div 1$	5	(93)	$24 \div 3$	8
(19)	$9 \div 1$	9	(44)	$30 \div 3$	10	(69)	$9 \div 3$	3	(94)	$14 \div 2$	7
(20)	$4 \div 2$	2	(45)	$8 \div 1$	8	(70)	$10 \div 1$	10	(95)	$32 \div 4$	8
(21)	$24 \div 3$	8	(46)	$15 \div 3$	5	(71)	$32 \div 4$	8	(96)	$16 \div 4$	4
(22)	$40 \div 4$	10	(47)	$16 \div 2$	8	(72)	$3 \div 3$	1	(97)	$15 \div 3$	5
(23)	$14 \div 2$	7	(48)	$16 \div 4$	4	(73)	$27 \div 3$	9	(98)	$10 \div 1$	10
(24)	$2 \div 1$	2	(49)	$28 \div 4$	7	(74)	$18 \div 3$	6	(99)	$18 \div 3$	6
(25)	$6 \div 3$	2	(50)	$4 \div 2$	2	(75)	$12 \div 2$	6	(100)	$6 \div 2$	3

Division 5

Name_____ Date _____ Score _____ Time _____

(1) $25 \div 5$		(26) $40 \div 5$		(51) $20 \div 5$		(76) $10 \div 5$	
(2) $10 \div 5$		(27) $5 \div 5$		(52) $25 \div 5$		(77) $45 \div 5$	
(3) $45 \div 5$		(28) $50 \div 5$		(53) $10 \div 5$		(78) $25 \div 5$	
(4) $5 \div 5$		(29) $30 \div 5$		(54) $5 \div 5$		(79) $15 \div 5$	
(5) $30 \div 5$		(30) $15 \div 5$		(55) $45 \div 5$		(80) $50 \div 5$	
(6) $40 \div 5$		(31) $20 \div 5$		(56) $20 \div 5$		(81) $5 \div 5$	
(7) $20 \div 5$		(32) $35 \div 5$		(57) $40 \div 5$		(82) $30 \div 5$	
(8) $50 \div 5$		(33) $40 \div 5$		(58) $15 \div 5$		(83) $45 \div 5$	
(9) $15 \div 5$		(34) $10 \div 5$		(59) $30 \div 5$		(84) $20 \div 5$	
(10) $35 \div 5$		(35) $15 \div 5$		(60) $25 \div 5$		(85) $40 \div 5$	
(11) $5 \div 5$		(36) $45 \div 5$		(61) $50 \div 5$		(86) $35 \div 5$	
(12) $40 \div 5$		(37) $25 \div 5$		(62) $5 \div 5$		(87) $10 \div 5$	
(13) $30 \div 5$		(38) $20 \div 5$		(63) $10 \div 5$		(88) $25 \div 5$	
(14) $20 \div 5$		(39) $50 \div 5$		(64) $35 \div 5$		(89) $15 \div 5$	
(15) $25 \div 5$		(40) $5 \div 5$		(65) $40 \div 5$		(90) $5 \div 5$	
(16) $10 \div 5$		(41) $30 \div 5$		(66) $15 \div 5$		(91) $30 \div 5$	
(17) $45 \div 5$		(42) $10 \div 5$		(67) $45 \div 5$		(92) $50 \div 5$	
(18) $40 \div 5$		(43) $15 \div 5$		(68) $25 \div 5$		(93) $25 \div 5$	
(19) $50 \div 5$		(44) $40 \div 5$		(69) $30 \div 5$		(94) $20 \div 5$	
(20) $30 \div 5$		(45) $35 \div 5$		(70) $10 \div 5$		(95) $45 \div 5$	
(21) $5 \div 5$		(46) $20 \div 5$		(71) $50 \div 5$		(96) $15 \div 5$	
(22) $35 \div 5$		(47) $10 \div 5$		(72) $45 \div 5$		(97) $40 \div 5$	
(23) $15 \div 5$		(48) $45 \div 5$		(73) $20 \div 5$		(98) $30 \div 5$	
(24) $10 \div 5$		(49) $25 \div 5$		(74) $5 \div 5$		(99) $50 \div 5$	
(25) $45 \div 5$		(50) $50 \div 5$		(75) $15 \div 5$		(100) $40 \div 5$	

Division 5

18

(1) 25 ÷ 5	5	(26) 40 ÷ 5	8	(51) 20 ÷ 5	4	(76) 10 ÷ 5	2
(2) 10 ÷ 5	2	(27) 5 ÷ 5	1	(52) 25 ÷ 5	5	(77) 45 ÷ 5	9
(3) 45 ÷ 5	9	(28) 50 ÷ 5	10	(53) 10 ÷ 5	2	(78) 25 ÷ 5	5
(4) 5 ÷ 5	1	(29) 30 ÷ 5	6	(54) 5 ÷ 5	1	(79) 15 ÷ 5	3
(5) 30 ÷ 5	6	(30) 15 ÷ 5	3	(55) 45 ÷ 5	9	(80) 50 ÷ 5	10
(6) 40 ÷ 5	8	(31) 20 ÷ 5	4	(56) 20 ÷ 5	4	(81) 5 ÷ 5	1
(7) 20 ÷ 5	4	(32) 35 ÷ 5	7	(57) 40 ÷ 5	8	(82) 30 ÷ 5	6
(8) 50 ÷ 5	10	(33) 40 ÷ 5	8	(58) 15 ÷ 5	3	(83) 45 ÷ 5	9
(9) 15 ÷ 5	3	(34) 10 ÷ 5	2	(59) 30 ÷ 5	6	(84) 20 ÷ 5	4
(10) 35 ÷ 5	7	(35) 15 ÷ 5	3	(60) 25 ÷ 5	5	(85) 40 ÷ 5	8
(11) 5 ÷ 5	1	(36) 45 ÷ 5	9	(61) 50 ÷ 5	10	(86) 35 ÷ 5	7
(12) 40 ÷ 5	8	(37) 25 ÷ 5	5	(62) 5 ÷ 5	1	(87) 10 ÷ 5	2
(13) 30 ÷ 5	6	(38) 20 ÷ 5	4	(63) 10 ÷ 5	2	(88) 25 ÷ 5	5
(14) 20 ÷ 5	4	(39) 50 ÷ 5	10	(64) 35 ÷ 5	7	(89) 15 ÷ 5	3
(15) 25 ÷ 5	5	(40) 5 ÷ 5	1	(65) 40 ÷ 5	8	(90) 5 ÷ 5	1
(16) 10 ÷ 5	2	(41) 30 ÷ 5	6	(66) 15 ÷ 5	3	(91) 30 ÷ 5	6
(17) 45 ÷ 5	9	(42) 10 ÷ 5	2	(67) 45 ÷ 5	9	(92) 50 ÷ 5	10
(18) 40 ÷ 5	8	(43) 15 ÷ 5	3	(68) 25 ÷ 5	5	(93) 25 ÷ 5	5
(19) 50 ÷ 5	10	(44) 40 ÷ 5	8	(69) 30 ÷ 5	6	(94) 20 ÷ 5	4
(20) 30 ÷ 5	6	(45) 35 ÷ 5	7	(70) 10 ÷ 5	2	(95) 45 ÷ 5	9
(21) 5 ÷ 5	1	(46) 20 ÷ 5	4	(71) 50 ÷ 5	10	(96) 15 ÷ 5	3
(22) 35 ÷ 5	7	(47) 10 ÷ 5	2	(72) 45 ÷ 5	9	(97) 40 ÷ 5	8
(23) 15 ÷ 5	3	(48) 45 ÷ 5	9	(73) 20 ÷ 5	4	(98) 30 ÷ 5	6
(24) 10 ÷ 5	2	(49) 25 ÷ 5	5	(74) 5 ÷ 5	1	(99) 50 ÷ 5	10
(25) 45 ÷ 5	9	(50) 50 ÷ 5	10	(75) 15 ÷ 5	3	(100) 40 ÷ 5	8

Division 1-5

Name_____ Date_____ Score_____ Time_____

(1) $24 \div 4$	(26) $12 \div 2$	(51) $20 \div 2$	(76) $1 \div 1$
(2) $6 \div 2$	(27) $8 \div 1$	(52) $25 \div 5$	(77) $12 \div 4$
(3) $9 \div 1$	(28) $8 \div 4$	(53) $14 \div 2$	(78) $30 \div 5$
(4) $15 \div 5$	(29) $30 \div 5$	(54) $10 \div 1$	(79) $40 \div 4$
(5) $4 \div 4$	(30) $32 \div 4$	(55) $16 \div 2$	(80) $8 \div 1$
(6) $24 \div 3$	(31) $1 \div 1$	(56) $27 \div 3$	(81) $5 \div 5$
(7) $5 \div 1$	(32) $30 \div 3$	(57) $16 \div 4$	(82) $12 \div 2$
(8) $45 \div 5$	(33) $12 \div 4$	(58) $50 \div 5$	(83) $8 \div 4$
(9) $18 \div 3$	(34) $4 \div 2$	(59) $4 \div 1$	(84) $32 \div 4$
(10) $2 \div 1$	(35) $40 \div 4$	(60) $6 \div 3$	(85) $30 \div 3$
(11) $36 \div 4$	(36) $7 \div 1$	(61) $20 \div 5$	(86) $4 \div 2$
(12) $20 \div 2$	(37) $5 \div 5$	(62) $40 \div 5$	(87) $7 \div 1$
(13) $12 \div 3$	(38) $15 \div 3$	(63) $2 \div 2$	(88) $15 \div 3$
(14) $25 \div 5$	(39) $28 \div 4$	(64) $12 \div 3$	(89) $8 \div 2$
(15) $2 \div 2$	(40) $8 \div 2$	(65) $36 \div 4$	(90) $9 \div 3$
(16) $14 \div 2$	(41) $10 \div 5$	(66) $2 \div 1$	(91) $35 \div 5$
(17) $40 \div 5$	(42) $9 \div 3$	(67) $18 \div 3$	(92) $3 \div 3$
(18) $10 \div 1$	(43) $6 \div 1$	(68) $45 \div 5$	(93) $18 \div 2$
(19) $20 \div 5$	(44) $35 \div 5$	(69) $5 \div 1$	(94) $21 \div 3$
(20) $16 \div 2$	(45) $20 \div 4$	(70) $24 \div 3$	(95) $3 \div 1$
(21) $6 \div 3$	(46) $3 \div 3$	(71) $4 \div 4$	(96) $10 \div 2$
(22) $27 \div 3$	(47) $10 \div 2$	(72) $15 \div 5$	(97) $20 \div 4$
(23) $4 \div 1$	(48) $3 \div 1$	(73) $9 \div 1$	(98) $6 \div 1$
(24) $16 \div 4$	(49) $18 \div 2$	(74) $6 \div 2$	(99) $10 \div 5$
(25) $50 \div 5$	(50) $21 \div 3$	(75) $24 \div 4$	(100) $28 \div 4$

Division 1-5

(1) 24 ÷ 4	6	(26) 12 ÷ 2	6	(51) 20 ÷ 2	10	(76) 1 ÷ 1	1
(2) 6 ÷ 2	3	(27) 8 ÷ 1	8	(52) 25 ÷ 5	5	(77) 12 ÷ 4	3
(3) 9 ÷ 1	9	(28) 8 ÷ 4	2	(53) 14 ÷ 2	7	(78) 30 ÷ 5	6
(4) 15 ÷ 5	3	(29) 30 ÷ 5	6	(54) 10 ÷ 1	10	(79) 40 ÷ 4	10
(5) 4 ÷ 4	1	(30) 32 ÷ 4	8	(55) 16 ÷ 2	8	(80) 8 ÷ 1	8
(6) 24 ÷ 3	8	(31) 1 ÷ 1	1	(56) 27 ÷ 3	9	(81) 5 ÷ 5	1
(7) 5 ÷ 1	5	(32) 30 ÷ 3	10	(57) 16 ÷ 4	4	(82) 12 ÷ 2	6
(8) 45 ÷ 5	9	(33) 12 ÷ 4	3	(58) 50 ÷ 5	10	(83) 8 ÷ 4	2
(9) 18 ÷ 3	6	(34) 4 ÷ 2	2	(59) 4 ÷ 1	4	(84) 32 ÷ 4	8
(10) 2 ÷ 1	2	(35) 40 ÷ 4	10	(60) 6 ÷ 3	2	(85) 30 ÷ 3	10
(11) 36 ÷ 4	9	(36) 7 ÷ 1	7	(61) 20 ÷ 5	4	(86) 4 ÷ 2	2
(12) 20 ÷ 2	10	(37) 5 ÷ 5	1	(62) 40 ÷ 5	8	(87) 7 ÷ 1	7
(13) 12 ÷ 3	4	(38) 15 ÷ 3	5	(63) 2 ÷ 2	1	(88) 15 ÷ 3	5
(14) 25 ÷ 5	5	(39) 28 ÷ 4	7	(64) 12 ÷ 3	4	(89) 8 ÷ 2	4
(15) 2 ÷ 2	1	(40) 8 ÷ 2	4	(65) 36 ÷ 4	9	(90) 9 ÷ 3	3
(16) 14 ÷ 2	7	(41) 10 ÷ 5	2	(66) 2 ÷ 1	2	(91) 35 ÷ 5	7
(17) 40 ÷ 5	8	(42) 9 ÷ 3	3	(67) 18 ÷ 3	6	(92) 3 ÷ 3	1
(18) 10 ÷ 1	10	(43) 6 ÷ 1	6	(68) 45 ÷ 5	9	(93) 18 ÷ 2	9
(19) 20 ÷ 5	4	(44) 35 ÷ 5	7	(69) 5 ÷ 1	5	(94) 21 ÷ 3	7
(20) 16 ÷ 2	8	(45) 20 ÷ 4	5	(70) 24 ÷ 3	8	(95) 3 ÷ 1	3
(21) 6 ÷ 3	2	(46) 3 ÷ 3	1	(71) 4 ÷ 4	1	(96) 10 ÷ 2	5
(22) 27 ÷ 3	9	(47) 10 ÷ 2	5	(72) 15 ÷ 5	3	(97) 20 ÷ 4	5
(23) 4 ÷ 1	4	(48) 3 ÷ 1	3	(73) 9 ÷ 1	9	(98) 6 ÷ 1	6
(24) 16 ÷ 4	4	(49) 18 ÷ 2	9	(74) 6 ÷ 2	3	(99) 10 ÷ 5	2
(25) 50 ÷ 5	10	(50) 21 ÷ 3	7	(75) 24 ÷ 4	6	(100) 28 ÷ 4	7

Division 6

Name_____ Date_____ Score_____ Time_____

(1) 60 ÷ 6		(26) 36 ÷ 6		(51) 42 ÷ 6		(76) 54 ÷ 6	
(2) 6 ÷ 6		(27) 6 ÷ 6		(52) 12 ÷ 6		(77) 60 ÷ 6	
(3) 18 ÷ 6		(28) 48 ÷ 6		(53) 36 ÷ 6		(78) 48 ÷ 6	
(4) 12 ÷ 6		(29) 54 ÷ 6		(54) 18 ÷ 6		(79) 30 ÷ 6	
(5) 30 ÷ 6		(30) 12 ÷ 6		(55) 60 ÷ 6		(80) 24 ÷ 6	
(6) 48 ÷ 6		(31) 18 ÷ 6		(56) 24 ÷ 6		(81) 54 ÷ 6	
(7) 12 ÷ 6		(32) 42 ÷ 6		(57) 48 ÷ 6		(82) 12 ÷ 6	
(8) 36 ÷ 6		(33) 6 ÷ 6		(58) 6 ÷ 6		(83) 42 ÷ 6	
(9) 24 ÷ 6		(34) 30 ÷ 6		(59) 54 ÷ 6		(84) 18 ÷ 6	
(10) 42 ÷ 6		(35) 24 ÷ 6		(60) 60 ÷ 6		(85) 48 ÷ 6	
(11) 12 ÷ 6		(36) 60 ÷ 6		(61) 12 ÷ 6		(86) 30 ÷ 6	
(12) 30 ÷ 6		(37) 12 ÷ 6		(62) 30 ÷ 6		(87) 36 ÷ 6	
(13) 54 ÷ 6		(38) 36 ÷ 6		(63) 18 ÷ 6		(88) 6 ÷ 6	
(14) 60 ÷ 6		(39) 42 ÷ 6		(64) 42 ÷ 6		(89) 60 ÷ 6	
(15) 18 ÷ 6		(40) 48 ÷ 6		(65) 48 ÷ 6		(90) 24 ÷ 6	
(16) 36 ÷ 6		(41) 18 ÷ 6		(66) 36 ÷ 6		(91) 54 ÷ 6	
(17) 6 ÷ 6		(42) 54 ÷ 6		(67) 60 ÷ 6		(92) 30 ÷ 6	
(18) 24 ÷ 6		(43) 36 ÷ 6		(68) 24 ÷ 6		(93) 18 ÷ 6	
(19) 48 ÷ 6		(44) 24 ÷ 6		(69) 42 ÷ 6		(94) 48 ÷ 6	
(20) 6 ÷ 6		(45) 42 ÷ 6		(70) 54 ÷ 6		(95) 36 ÷ 6	
(21) 12 ÷ 6		(46) 6 ÷ 6		(71) 30 ÷ 6		(96) 12 ÷ 6	
(22) 30 ÷ 6		(47) 60 ÷ 6		(72) 6 ÷ 6		(97) 42 ÷ 6	
(23) 54 ÷ 6		(48) 30 ÷ 6		(73) 48 ÷ 6		(98) 6 ÷ 6	
(24) 36 ÷ 6		(49) 18 ÷ 6		(74) 18 ÷ 6		(99) 60 ÷ 6	
(25) 42 ÷ 6		(50) 54 ÷ 6		(75) 24 ÷ 6		(100) 24 ÷ 6	

Division 6

(1) 60 ÷ 6	10	(26) 36 ÷ 6	6	(51) 42 ÷ 6	7	(76) 54 ÷ 6	9
(2) 6 ÷ 6	1	(27) 6 ÷ 6	1	(52) 12 ÷ 6	2	(77) 60 ÷ 6	10
(3) 18 ÷ 6	3	(28) 48 ÷ 6	8	(53) 36 ÷ 6	6	(78) 48 ÷ 6	8
(4) 12 ÷ 6	2	(29) 54 ÷ 6	9	(54) 18 ÷ 6	3	(79) 30 ÷ 6	5
(5) 30 ÷ 6	5	(30) 12 ÷ 6	2	(55) 60 ÷ 6	10	(80) 24 ÷ 6	4
(6) 48 ÷ 6	8	(31) 18 ÷ 6	3	(56) 24 ÷ 6	4	(81) 54 ÷ 6	9
(7) 12 ÷ 6	2	(32) 42 ÷ 6	7	(57) 48 ÷ 6	8	(82) 12 ÷ 6	2
(8) 36 ÷ 6	6	(33) 6 ÷ 6	1	(58) 6 ÷ 6	1	(83) 42 ÷ 6	7
(9) 24 ÷ 6	4	(34) 30 ÷ 6	5	(59) 54 ÷ 6	9	(84) 18 ÷ 6	3
(10) 42 ÷ 6	7	(35) 24 ÷ 6	4	(60) 60 ÷ 6	10	(85) 48 ÷ 6	8
(11) 12 ÷ 6	2	(36) 60 ÷ 6	10	(61) 12 ÷ 6	2	(86) 30 ÷ 6	5
(12) 30 ÷ 6	5	(37) 12 ÷ 6	2	(62) 30 ÷ 6	5	(87) 36 ÷ 6	6
(13) 54 ÷ 6	9	(38) 36 ÷ 6	6	(63) 18 ÷ 6	3	(88) 6 ÷ 6	1
(14) 60 ÷ 6	10	(39) 42 ÷ 6	7	(64) 42 ÷ 6	7	(89) 60 ÷ 6	10
(15) 18 ÷ 6	3	(40) 48 ÷ 6	8	(65) 48 ÷ 6	8	(90) 24 ÷ 6	4
(16) 36 ÷ 6	6	(41) 18 ÷ 6	3	(66) 36 ÷ 6	6	(91) 54 ÷ 6	9
(17) 6 ÷ 6	1	(42) 54 ÷ 6	9	(67) 60 ÷ 6	10	(92) 30 ÷ 6	5
(18) 24 ÷ 6	4	(43) 36 ÷ 6	6	(68) 24 ÷ 6	4	(93) 18 ÷ 6	3
(19) 48 ÷ 6	8	(44) 24 ÷ 6	4	(69) 42 ÷ 6	7	(94) 48 ÷ 6	8
(20) 6 ÷ 6	1	(45) 42 ÷ 6	7	(70) 54 ÷ 6	9	(95) 36 ÷ 6	6
(21) 12 ÷ 6	2	(46) 6 ÷ 6	1	(71) 30 ÷ 6	5	(96) 12 ÷ 6	2
(22) 30 ÷ 6	5	(47) 60 ÷ 6	10	(72) 6 ÷ 6	1	(97) 42 ÷ 6	7
(23) 54 ÷ 6	9	(48) 30 ÷ 6	5	(73) 48 ÷ 6	8	(98) 6 ÷ 6	1
(24) 36 ÷ 6	6	(49) 18 ÷ 6	3	(74) 18 ÷ 6	3	(99) 60 ÷ 6	10
(25) 42 ÷ 6	7	(50) 54 ÷ 6	9	(75) 24 ÷ 6	4	(100) 24 ÷ 6	4

Division 1-6

Name_____ Date_____ Score_____ Time_____

(1) $8 \div 4$	(26) $20 \div 2$	(51) $36 \div 6$	(76) $54 \div 6$
(2) $30 \div 6$	(27) $16 \div 4$	(52) $28 \div 4$	(77) $36 \div 4$
(3) $12 \div 3$	(28) $18 \div 3$	(53) $16 \div 2$	(78) $18 \div 2$
(4) $18 \div 2$	(29) $5 \div 5$	(54) $3 \div 1$	(79) $35 \div 5$
(5) $6 \div 1$	(30) $12 \div 4$	(55) $45 \div 5$	(80) $30 \div 6$
(6) $35 \div 5$	(31) $27 \div 3$	(56) $20 \div 4$	(81) $18 \div 3$
(7) $54 \div 6$	(32) $18 \div 6$	(57) $30 \div 3$	(82) $8 \div 4$
(8) $4 \div 2$	(33) $4 \div 1$	(58) $10 \div 2$	(83) $20 \div 2$
(9) $36 \div 4$	(34) $40 \div 5$	(59) $2 \div 2$	(84) $5 \div 5$
(10) $12 \div 6$	(35) $60 \div 6$	(60) $15 \div 3$	(85) $27 \div 3$
(11) $2 \div 1$	(36) $7 \div 1$	(61) $48 \div 6$	(86) $18 \div 6$
(12) $14 \div 2$	(37) $4 \div 4$	(62) $1 \div 1$	(87) $12 \div 4$
(13) $10 \div 5$	(38) $25 \div 5$	(63) $6 \div 3$	(88) $40 \div 5$
(14) $24 \div 4$	(39) $24 \div 6$	(64) $20 \div 5$	(89) $4 \div 1$
(15) $42 \div 6$	(40) $12 \div 2$	(65) $5 \div 1$	(90) $25 \div 5$
(16) $8 \div 1$	(41) $40 \div 4$	(66) $8 \div 2$	(91) $45 \div 5$
(17) $50 \div 5$	(42) $6 \div 6$	(67) $24 \div 3$	(92) $36 \div 6$
(18) $10 \div 1$	(43) $9 \div 3$	(68) $10 \div 1$	(93) $21 \div 3$
(19) $24 \div 3$	(44) $32 \div 4$	(69) $50 \div 5$	(94) $24 \div 6$
(20) $8 \div 2$	(45) $15 \div 5$	(70) $8 \div 1$	(95) $10 \div 1$
(21) $5 \div 1$	(46) $9 \div 1$	(71) $42 \div 6$	(96) $48 \div 6$
(22) $20 \div 5$	(47) $30 \div 5$	(72) $24 \div 4$	(97) $12 \div 2$
(23) $6 \div 3$	(48) $3 \div 3$	(73) $10 \div 5$	(98) $4 \div 4$
(24) $1 \div 1$	(49) $21 \div 3$	(74) $14 \div 2$	(99) $60 \div 6$
(25) $48 \div 6$	(50) $6 \div 2$	(75) $12 \div 6$	(100) $15 \div 5$

24

Division 1-6

(1) 8 ÷ 4	2	(26) 20 ÷ 2	10	(51) 36 ÷ 6	6	(76) 54 ÷ 6	9
(2) 30 ÷ 6	5	(27) 16 ÷ 4	4	(52) 28 ÷ 4	7	(77) 36 ÷ 4	9
(3) 12 ÷ 3	4	(28) 18 ÷ 3	6	(53) 16 ÷ 2	8	(78) 18 ÷ 2	9
(4) 18 ÷ 2	9	(29) 5 ÷ 5	1	(54) 3 ÷ 1	3	(79) 35 ÷ 5	7
(5) 6 ÷ 1	6	(30) 12 ÷ 4	3	(55) 45 ÷ 5	9	(80) 30 ÷ 6	5
(6) 35 ÷ 5	7	(31) 27 ÷ 3	9	(56) 20 ÷ 4	5	(81) 18 ÷ 3	6
(7) 54 ÷ 6	9	(32) 18 ÷ 6	3	(57) 30 ÷ 3	10	(82) 8 ÷ 4	2
(8) 4 ÷ 2	2	(33) 4 ÷ 1	4	(58) 10 ÷ 2	5	(83) 20 ÷ 2	10
(9) 36 ÷ 4	9	(34) 40 ÷ 5	8	(59) 2 ÷ 2	1	(84) 5 ÷ 5	1
(10) 12 ÷ 6	2	(35) 60 ÷ 6	10	(60) 15 ÷ 3	5	(85) 27 ÷ 3	9
(11) 2 ÷ 1	2	(36) 7 ÷ 1	7	(61) 48 ÷ 6	8	(86) 18 ÷ 6	3
(12) 14 ÷ 2	7	(37) 4 ÷ 4	1	(62) 1 ÷ 1	1	(87) 12 ÷ 4	3
(13) 10 ÷ 5	2	(38) 25 ÷ 5	5	(63) 6 ÷ 3	2	(88) 40 ÷ 5	8
(14) 24 ÷ 4	6	(39) 24 ÷ 6	4	(64) 20 ÷ 5	4	(89) 4 ÷ 1	4
(15) 42 ÷ 6	7	(40) 12 ÷ 2	6	(65) 5 ÷ 1	5	(90) 25 ÷ 5	5
(16) 8 ÷ 1	8	(41) 40 ÷ 4	10	(66) 8 ÷ 2	4	(91) 45 ÷ 5	9
(17) 50 ÷ 5	10	(42) 6 ÷ 6	1	(67) 24 ÷ 3	8	(92) 36 ÷ 6	6
(18) 10 ÷ 1	10	(43) 9 ÷ 3	3	(68) 10 ÷ 1	10	(93) 21 ÷ 3	7
(19) 24 ÷ 3	8	(44) 32 ÷ 4	8	(69) 50 ÷ 5	10	(94) 24 ÷ 6	4
(20) 8 ÷ 2	4	(45) 15 ÷ 5	3	(70) 8 ÷ 1	8	(95) 10 ÷ 1	10
(21) 5 ÷ 1	5	(46) 9 ÷ 1	9	(71) 42 ÷ 6	7	(96) 48 ÷ 6	8
(22) 20 ÷ 5	4	(47) 30 ÷ 5	6	(72) 24 ÷ 4	6	(97) 12 ÷ 2	6
(23) 6 ÷ 3	2	(48) 3 ÷ 3	1	(73) 10 ÷ 5	2	(98) 4 ÷ 4	1
(24) 1 ÷ 1	1	(49) 21 ÷ 3	7	(74) 14 ÷ 2	7	(99) 60 ÷ 6	10
(25) 48 ÷ 6	8	(50) 6 ÷ 2	3	(75) 12 ÷ 6	2	(100) 15 ÷ 5	3

Division 7

Name_____ Date _____ Score _____ Time _____

(1) $28 \div 7$	(26) $49 \div 7$	(51) $56 \div 7$	(76) $7 \div 7$	
(2) $7 \div 7$	(27) $14 \div 7$	(52) $70 \div 7$	(77) $42 \div 7$	
(3) $56 \div 7$	(28) $42 \div 7$	(53) $35 \div 7$	(78) $56 \div 7$	
(4) $35 \div 7$	(29) $21 \div 7$	(54) $49 \div 7$	(79) $21 \div 7$	
(5) $49 \div 7$	(30) $70 \div 7$	(55) $21 \div 7$	(80) $35 \div 7$	
(6) $21 \div 7$	(31) $28 \div 7$	(56) $56 \div 7$	(81) $63 \div 7$	
(7) $7 \div 7$	(32) $63 \div 7$	(57) $28 \div 7$	(82) $28 \div 7$	
(8) $70 \div 7$	(33) $42 \div 7$	(58) $42 \div 7$	(83) $49 \div 7$	
(9) $14 \div 7$	(34) $35 \div 7$	(59) $63 \div 7$	(84) $35 \div 7$	
(10) $42 \div 7$	(35) $7 \div 7$	(60) $14 \div 7$	(85) $21 \div 7$	
(11) $63 \div 7$	(36) $49 \div 7$	(61) $21 \div 7$	(86) $70 \div 7$	
(12) $7 \div 7$	(37) $56 \div 7$	(62) $28 \div 7$	(87) $14 \div 7$	
(13) $28 \div 7$	(38) $21 \div 7$	(63) $7 \div 7$	(88) $42 \div 7$	
(14) $49 \div 7$	(39) $14 \div 7$	(64) $49 \div 7$	(89) $28 \div 7$	
(15) $21 \div 7$	(40) $70 \div 7$	(65) $42 \div 7$	(90) $63 \div 7$	
(16) $56 \div 7$	(41) $49 \div 7$	(66) $56 \div 7$	(91) $7 \div 7$	
(17) $63 \div 7$	(42) $42 \div 7$	(67) $35 \div 7$	(92) $56 \div 7$	
(18) $49 \div 7$	(43) $63 \div 7$	(68) $70 \div 7$	(93) $35 \div 7$	
(19) $35 \div 7$	(44) $28 \div 7$	(69) $49 \div 7$	(94) $49 \div 7$	
(20) $70 \div 7$	(45) $7 \div 7$	(70) $63 \div 7$	(95) $28 \div 7$	
(21) $14 \div 7$	(46) $56 \div 7$	(71) $7 \div 7$	(96) $7 \div 7$	
(22) $42 \div 7$	(47) $35 \div 7$	(72) $21 \div 7$	(97) $14 \div 7$	
(23) $63 \div 7$	(48) $21 \div 7$	(73) $14 \div 7$	(98) $70 \div 7$	
(24) $28 \div 7$	(49) $42 \div 7$	(74) $70 \div 7$	(99) $35 \div 7$	
(25) $70 \div 7$	(50) $14 \div 7$	(75) $56 \div 7$	(100) $63 \div 7$	

Division 7

(1) 28 ÷ 7 = **4**	(26) 49 ÷ 7 = **7**	(51) 56 ÷ 7 = **8**	(76) 7 ÷ 7 = **1**	
(2) 7 ÷ 7 = **1**	(27) 14 ÷ 7 = **2**	(52) 70 ÷ 7 = **10**	(77) 42 ÷ 7 = **6**	
(3) 56 ÷ 7 = **8**	(28) 42 ÷ 7 = **6**	(53) 35 ÷ 7 = **5**	(78) 56 ÷ 7 = **8**	
(4) 35 ÷ 7 = **5**	(29) 21 ÷ 7 = **3**	(54) 49 ÷ 7 = **7**	(79) 21 ÷ 7 = **3**	
(5) 49 ÷ 7 = **7**	(30) 70 ÷ 7 = **10**	(55) 21 ÷ 7 = **3**	(80) 35 ÷ 7 = **5**	
(6) 21 ÷ 7 = **3**	(31) 28 ÷ 7 = **4**	(56) 56 ÷ 7 = **8**	(81) 63 ÷ 7 = **9**	
(7) 7 ÷ 7 = **1**	(32) 63 ÷ 7 = **9**	(57) 28 ÷ 7 = **4**	(82) 28 ÷ 7 = **4**	
(8) 70 ÷ 7 = **10**	(33) 42 ÷ 7 = **6**	(58) 42 ÷ 7 = **6**	(83) 49 ÷ 7 = **7**	
(9) 14 ÷ 7 = **2**	(34) 35 ÷ 7 = **5**	(59) 63 ÷ 7 = **9**	(84) 35 ÷ 7 = **5**	
(10) 42 ÷ 7 = **6**	(35) 7 ÷ 7 = **1**	(60) 14 ÷ 7 = **2**	(85) 21 ÷ 7 = **3**	
(11) 63 ÷ 7 = **9**	(36) 49 ÷ 7 = **7**	(61) 21 ÷ 7 = **3**	(86) 70 ÷ 7 = **10**	
(12) 7 ÷ 7 = **1**	(37) 56 ÷ 7 = **8**	(62) 28 ÷ 7 = **4**	(87) 14 ÷ 7 = **2**	
(13) 28 ÷ 7 = **4**	(38) 21 ÷ 7 = **3**	(63) 7 ÷ 7 = **1**	(88) 42 ÷ 7 = **6**	
(14) 49 ÷ 7 = **7**	(39) 14 ÷ 7 = **2**	(64) 49 ÷ 7 = **7**	(89) 28 ÷ 7 = **4**	
(15) 21 ÷ 7 = **3**	(40) 70 ÷ 7 = **10**	(65) 42 ÷ 7 = **6**	(90) 63 ÷ 7 = **9**	
(16) 56 ÷ 7 = **8**	(41) 49 ÷ 7 = **7**	(66) 56 ÷ 7 = **8**	(91) 7 ÷ 7 = **1**	
(17) 63 ÷ 7 = **9**	(42) 42 ÷ 7 = **6**	(67) 35 ÷ 7 = **5**	(92) 56 ÷ 7 = **8**	
(18) 49 ÷ 7 = **7**	(43) 63 ÷ 7 = **9**	(68) 70 ÷ 7 = **10**	(93) 35 ÷ 7 = **5**	
(19) 35 ÷ 7 = **5**	(44) 28 ÷ 7 = **4**	(69) 49 ÷ 7 = **7**	(94) 49 ÷ 7 = **7**	
(20) 70 ÷ 7 = **10**	(45) 7 ÷ 7 = **1**	(70) 63 ÷ 7 = **9**	(95) 28 ÷ 7 = **4**	
(21) 14 ÷ 7 = **2**	(46) 56 ÷ 7 = **8**	(71) 7 ÷ 7 = **1**	(96) 7 ÷ 7 = **1**	
(22) 42 ÷ 7 = **6**	(47) 35 ÷ 7 = **5**	(72) 21 ÷ 7 = **3**	(97) 14 ÷ 7 = **2**	
(23) 63 ÷ 7 = **9**	(48) 21 ÷ 7 = **3**	(73) 14 ÷ 7 = **2**	(98) 70 ÷ 7 = **10**	
(24) 28 ÷ 7 = **4**	(49) 42 ÷ 7 = **6**	(74) 70 ÷ 7 = **10**	(99) 35 ÷ 7 = **5**	
(25) 70 ÷ 7 = **10**	(50) 14 ÷ 7 = **2**	(75) 56 ÷ 7 = **8**	(100) 63 ÷ 7 = **9**	

Division 1-7

Name_____ Date _____ Score _____ Time _____

(1) $40 \div 5$		(26) $56 \div 7$		(51) $30 \div 6$		(76) $3 \div 3$	
(2) $6 \div 2$		(27) $60 \div 6$		(52) $4 \div 2$		(77) $36 \div 6$	
(3) $63 \div 7$		(28) $9 \div 3$		(53) $10 \div 5$		(78) $32 \div 4$	
(4) $18 \div 6$		(29) $24 \div 4$		(54) $49 \div 7$		(79) $30 \div 5$	
(5) $27 \div 3$		(30) $5 \div 5$		(55) $3 \div 1$		(80) $35 \div 7$	
(6) $8 \div 1$		(31) $12 \div 6$		(56) $28 \div 7$		(81) $8 \div 4$	
(7) $20 \div 4$		(32) $18 \div 2$		(57) $12 \div 3$		(82) $12 \div 2$	
(8) $14 \div 7$		(33) $21 \div 7$		(58) $7 \div 1$		(83) $28 \div 4$	
(9) $20 \div 2$		(34) $35 \div 5$		(59) $24 \div 3$		(84) $56 \div 7$	
(10) $42 \div 7$		(35) $6 \div 1$		(60) $6 \div 6$		(85) $9 \div 3$	
(11) $3 \div 3$		(36) $7 \div 7$		(61) $36 \div 4$		(86) $5 \div 5$	
(12) $36 \div 6$		(37) $54 \div 6$		(62) $10 \div 2$		(87) $18 \div 2$	
(13) $4 \div 1$		(38) $40 \div 4$		(63) $16 \div 4$		(88) $35 \div 5$	
(14) $32 \div 4$		(39) $2 \div 1$		(64) $14 \div 2$		(89) $7 \div 7$	
(15) $15 \div 3$		(40) $24 \div 6$		(65) $45 \div 5$		(90) $40 \div 4$	
(16) $30 \div 5$		(41) $15 \div 5$		(66) $21 \div 3$		(91) $24 \div 6$	
(17) $2 \div 2$		(42) $70 \div 7$		(67) $9 \div 1$		(92) $8 \div 2$	
(18) $35 \div 7$		(43) $8 \div 2$		(68) $30 \div 3$		(93) $42 \div 6$	
(19) $48 \div 6$		(44) $5 \div 1$		(69) $25 \div 5$		(94) $18 \div 3$	
(20) $8 \div 4$		(45) $42 \div 6$		(70) $4 \div 4$		(95) $30 \div 6$	
(21) $20 \div 5$		(46) $12 \div 4$		(71) $40 \div 5$		(96) $10 \div 5$	
(22) $12 \div 2$		(47) $1 \div 1$		(72) $63 \div 7$		(97) $18 \div 6$	
(23) $10 \div 1$		(48) $16 \div 2$		(73) $27 \div 3$		(98) $49 \div 7$	
(24) $28 \div 4$		(49) $18 \div 3$		(74) $20 \div 4$		(99) $14 \div 7$	
(25) $50 \div 5$		(50) $6 \div 3$		(75) $20 \div 2$		(100) $15 \div 3$	

28

Division 1-7

(1) 40 ÷ 5	8	(26) 56 ÷ 7	8	(51) 30 ÷ 6	5	(76) 3 ÷ 3	1
(2) 6 ÷ 2	3	(27) 60 ÷ 6	10	(52) 4 ÷ 2	2	(77) 36 ÷ 6	6
(3) 63 ÷ 7	9	(28) 9 ÷ 3	3	(53) 10 ÷ 5	2	(78) 32 ÷ 4	8
(4) 18 ÷ 6	3	(29) 24 ÷ 4	6	(54) 49 ÷ 7	7	(79) 30 ÷ 5	6
(5) 27 ÷ 3	9	(30) 5 ÷ 5	1	(55) 3 ÷ 1	3	(80) 35 ÷ 7	5
(6) 8 ÷ 1	8	(31) 12 ÷ 6	2	(56) 28 ÷ 7	4	(81) 8 ÷ 4	2
(7) 20 ÷ 4	5	(32) 18 ÷ 2	9	(57) 12 ÷ 3	4	(82) 12 ÷ 2	6
(8) 14 ÷ 7	2	(33) 21 ÷ 7	3	(58) 7 ÷ 1	7	(83) 28 ÷ 4	7
(9) 20 ÷ 2	10	(34) 35 ÷ 5	7	(59) 24 ÷ 3	8	(84) 56 ÷ 7	8
(10) 42 ÷ 7	6	(35) 6 ÷ 1	6	(60) 6 ÷ 6	1	(85) 9 ÷ 3	3
(11) 3 ÷ 3	1	(36) 7 ÷ 7	1	(61) 36 ÷ 4	9	(86) 5 ÷ 5	1
(12) 36 ÷ 6	6	(37) 54 ÷ 6	9	(62) 10 ÷ 2	5	(87) 18 ÷ 2	9
(13) 4 ÷ 1	4	(38) 40 ÷ 4	10	(63) 16 ÷ 4	4	(88) 35 ÷ 5	7
(14) 32 ÷ 4	8	(39) 2 ÷ 1	2	(64) 14 ÷ 2	7	(89) 7 ÷ 7	1
(15) 15 ÷ 3	5	(40) 24 ÷ 6	4	(65) 45 ÷ 5	9	(90) 40 ÷ 4	10
(16) 30 ÷ 5	6	(41) 15 ÷ 5	3	(66) 21 ÷ 3	7	(91) 24 ÷ 6	4
(17) 2 ÷ 2	1	(42) 70 ÷ 7	10	(67) 9 ÷ 1	9	(92) 8 ÷ 2	4
(18) 35 ÷ 7	5	(43) 8 ÷ 2	4	(68) 30 ÷ 3	10	(93) 42 ÷ 6	7
(19) 48 ÷ 6	8	(44) 5 ÷ 1	5	(69) 25 ÷ 5	5	(94) 18 ÷ 3	6
(20) 8 ÷ 4	2	(45) 42 ÷ 6	7	(70) 4 ÷ 4	1	(95) 30 ÷ 6	5
(21) 20 ÷ 5	4	(46) 12 ÷ 4	3	(71) 40 ÷ 5	8	(96) 10 ÷ 5	2
(22) 12 ÷ 2	6	(47) 1 ÷ 1	1	(72) 63 ÷ 7	9	(97) 18 ÷ 6	3
(23) 10 ÷ 1	10	(48) 16 ÷ 2	8	(73) 27 ÷ 3	9	(98) 49 ÷ 7	7
(24) 28 ÷ 4	7	(49) 18 ÷ 3	6	(74) 20 ÷ 4	5	(99) 14 ÷ 7	2
(25) 50 ÷ 5	10	(50) 6 ÷ 3	2	(75) 20 ÷ 2	10	(100) 15 ÷ 3	5

Division 8

Name_____ Date_____ Score_____ Time_____

(1) 48 ÷ 8	(26) 80 ÷ 8	(51) 64 ÷ 8	(76) 56 ÷ 8
(2) 24 ÷ 8	(27) 56 ÷ 8	(52) 16 ÷ 8	(77) 72 ÷ 8
(3) 72 ÷ 8	(28) 32 ÷ 8	(53) 72 ÷ 8	(78) 24 ÷ 8
(4) 56 ÷ 8	(29) 8 ÷ 8	(54) 8 ÷ 8	(79) 80 ÷ 8
(5) 8 ÷ 8	(30) 24 ÷ 8	(55) 40 ÷ 8	(80) 16 ÷ 8
(6) 40 ÷ 8	(31) 16 ÷ 8	(56) 32 ÷ 8	(81) 64 ÷ 8
(7) 64 ÷ 8	(32) 72 ÷ 8	(57) 56 ÷ 8	(82) 48 ÷ 8
(8) 32 ÷ 8	(33) 48 ÷ 8	(58) 24 ÷ 8	(83) 8 ÷ 8
(9) 80 ÷ 8	(34) 8 ÷ 8	(59) 16 ÷ 8	(84) 40 ÷ 8
(10) 16 ÷ 8	(35) 64 ÷ 8	(60) 80 ÷ 8	(85) 32 ÷ 8
(11) 72 ÷ 8	(36) 40 ÷ 8	(61) 8 ÷ 8	(86) 72 ÷ 8
(12) 56 ÷ 8	(37) 80 ÷ 8	(62) 48 ÷ 8	(87) 16 ÷ 8
(13) 24 ÷ 8	(38) 16 ÷ 8	(63) 64 ÷ 8	(88) 80 ÷ 8
(14) 40 ÷ 8	(39) 32 ÷ 8	(64) 40 ÷ 8	(89) 56 ÷ 8
(15) 48 ÷ 8	(40) 72 ÷ 8	(65) 16 ÷ 8	(90) 32 ÷ 8
(16) 8 ÷ 8	(41) 56 ÷ 8	(66) 80 ÷ 8	(91) 48 ÷ 8
(17) 72 ÷ 8	(42) 24 ÷ 8	(67) 32 ÷ 8	(92) 24 ÷ 8
(18) 32 ÷ 8	(43) 16 ÷ 8	(68) 48 ÷ 8	(93) 64 ÷ 8
(19) 64 ÷ 8	(44) 48 ÷ 8	(69) 72 ÷ 8	(94) 8 ÷ 8
(20) 80 ÷ 8	(45) 8 ÷ 8	(70) 24 ÷ 8	(95) 40 ÷ 8
(21) 16 ÷ 8	(46) 40 ÷ 8	(71) 56 ÷ 8	(96) 72 ÷ 8
(22) 56 ÷ 8	(47) 32 ÷ 8	(72) 80 ÷ 8	(97) 32 ÷ 8
(23) 40 ÷ 8	(48) 64 ÷ 8	(73) 64 ÷ 8	(98) 24 ÷ 8
(24) 64 ÷ 8	(49) 48 ÷ 8	(74) 40 ÷ 8	(99) 80 ÷ 8
(25) 8 ÷ 8	(50) 24 ÷ 8	(75) 48 ÷ 8	(100) 56 ÷ 8

Division 8

(1) $48 \div 8$	6	
(2) $24 \div 8$	3	
(3) $72 \div 8$	9	
(4) $56 \div 8$	7	
(5) $8 \div 8$	1	
(6) $40 \div 8$	5	
(7) $64 \div 8$	8	
(8) $32 \div 8$	4	
(9) $80 \div 8$	10	
(10) $16 \div 8$	2	
(11) $72 \div 8$	9	
(12) $56 \div 8$	7	
(13) $24 \div 8$	3	
(14) $40 \div 8$	5	
(15) $48 \div 8$	6	
(16) $8 \div 8$	1	
(17) $72 \div 8$	9	
(18) $32 \div 8$	4	
(19) $64 \div 8$	8	
(20) $80 \div 8$	10	
(21) $16 \div 8$	2	
(22) $56 \div 8$	7	
(23) $40 \div 8$	5	
(24) $64 \div 8$	8	
(25) $8 \div 8$	1	

(26) $80 \div 8$	10	
(27) $56 \div 8$	7	
(28) $32 \div 8$	4	
(29) $8 \div 8$	1	
(30) $24 \div 8$	3	
(31) $16 \div 8$	2	
(32) $72 \div 8$	9	
(33) $48 \div 8$	6	
(34) $8 \div 8$	1	
(35) $64 \div 8$	8	
(36) $40 \div 8$	5	
(37) $80 \div 8$	10	
(38) $16 \div 8$	2	
(39) $32 \div 8$	4	
(40) $72 \div 8$	9	
(41) $56 \div 8$	7	
(42) $24 \div 8$	3	
(43) $16 \div 8$	2	
(44) $48 \div 8$	6	
(45) $8 \div 8$	1	
(46) $40 \div 8$	5	
(47) $32 \div 8$	4	
(48) $64 \div 8$	8	
(49) $48 \div 8$	6	
(50) $24 \div 8$	3	

(51) $64 \div 8$	8	
(52) $16 \div 8$	2	
(53) $72 \div 8$	9	
(54) $8 \div 8$	1	
(55) $40 \div 8$	5	
(56) $32 \div 8$	4	
(57) $56 \div 8$	7	
(58) $24 \div 8$	3	
(59) $16 \div 8$	2	
(60) $80 \div 8$	10	
(61) $8 \div 8$	1	
(62) $48 \div 8$	6	
(63) $64 \div 8$	8	
(64) $40 \div 8$	5	
(65) $16 \div 8$	2	
(66) $80 \div 8$	10	
(67) $32 \div 8$	4	
(68) $48 \div 8$	6	
(69) $72 \div 8$	9	
(70) $24 \div 8$	3	
(71) $56 \div 8$	7	
(72) $80 \div 8$	10	
(73) $64 \div 8$	8	
(74) $40 \div 8$	5	
(75) $48 \div 8$	6	

(76) $56 \div 8$	7	
(77) $72 \div 8$	9	
(78) $24 \div 8$	3	
(79) $80 \div 8$	10	
(80) $16 \div 8$	2	
(81) $64 \div 8$	8	
(82) $48 \div 8$	6	
(83) $8 \div 8$	1	
(84) $40 \div 8$	5	
(85) $32 \div 8$	4	
(86) $72 \div 8$	9	
(87) $16 \div 8$	2	
(88) $80 \div 8$	10	
(89) $56 \div 8$	7	
(90) $32 \div 8$	4	
(91) $48 \div 8$	6	
(92) $24 \div 8$	3	
(93) $64 \div 8$	8	
(94) $8 \div 8$	1	
(95) $40 \div 8$	5	
(96) $72 \div 8$	9	
(97) $32 \div 8$	4	
(98) $24 \div 8$	3	
(99) $80 \div 8$	10	
(100) $56 \div 8$	7	

Division 1-8

Name _____ Date _____ Score _____ Time _____

(1) $28 \div 7$	(26) $8 \div 2$	(51) $54 \div 6$	(76) $4 \div 2$
(2) $14 \div 2$	(27) $5 \div 5$	(52) $12 \div 2$	(77) $36 \div 4$
(3) $24 \div 4$	(28) $32 \div 4$	(53) $30 \div 5$	(78) $60 \div 6$
(4) $50 \div 5$	(29) $9 \div 3$	(54) $64 \div 8$	(79) $6 \div 1$
(5) $3 \div 1$	(30) $56 \div 8$	(55) $6 \div 2$	(80) $36 \div 6$
(6) $42 \div 6$	(31) $25 \div 5$	(56) $70 \div 7$	(81) $28 \div 7$
(7) $6 \div 3$	(32) $1 \div 1$	(57) $12 \div 4$	(82) $24 \div 4$
(8) $10 \div 1$	(33) $14 \div 7$	(58) $4 \div 1$	(83) $3 \div 1$
(9) $24 \div 3$	(34) $30 \div 3$	(59) $30 \div 6$	(84) $24 \div 3$
(10) $20 \div 5$	(35) $7 \div 1$	(60) $40 \div 8$	(85) $63 \div 7$
(11) $6 \div 6$	(36) $16 \div 4$	(61) $7 \div 7$	(86) $32 \div 8$
(12) $63 \div 7$	(37) $72 \div 8$	(62) $10 \div 2$	(87) $56 \div 7$
(13) $48 \div 8$	(38) $10 \div 5$	(63) $49 \div 7$	(88) $40 \div 5$
(14) $8 \div 4$	(39) $9 \div 1$	(64) $20 \div 4$	(89) $42 \div 7$
(15) $32 \div 8$	(40) $28 \div 4$	(65) $2 \div 1$	(90) $25 \div 5$
(16) $18 \div 3$	(41) $12 \div 6$	(66) $27 \div 3$	(91) $24 \div 6$
(17) $16 \div 2$	(42) $15 \div 3$	(67) $21 \div 7$	(92) $14 \div 7$
(18) $56 \div 7$	(43) $8 \div 1$	(68) $45 \div 5$	(93) $28 \div 4$
(19) $24 \div 8$	(44) $35 \div 5$	(69) $20 \div 2$	(94) $64 \div 8$
(20) $40 \div 5$	(45) $18 \div 6$	(70) $4 \div 4$	(95) $35 \div 5$
(21) $5 \div 1$	(46) $12 \div 3$	(71) $35 \div 7$	(96) $40 \div 8$
(22) $2 \div 2$	(47) $40 \div 4$	(72) $48 \div 6$	(97) $72 \div 8$
(23) $42 \div 7$	(48) $16 \div 8$	(73) $21 \div 3$	(98) $49 \div 7$
(24) $80 \div 8$	(49) $3 \div 3$	(74) $15 \div 5$	(99) $64 \div 8$
(25) $24 \div 6$	(50) $18 \div 2$	(75) $8 \div 8$	(100) $18 \div 6$

Division 1-8

(1) 28 ÷ 7	4	(26) 8 ÷ 2	4	(51) 54 ÷ 6	9	(76) 4 ÷ 2	2
(2) 14 ÷ 2	7	(27) 5 ÷ 5	1	(52) 12 ÷ 2	6	(77) 36 ÷ 4	9
(3) 24 ÷ 4	6	(28) 32 ÷ 4	8	(53) 30 ÷ 5	6	(78) 60 ÷ 6	10
(4) 50 ÷ 5	10	(29) 9 ÷ 3	3	(54) 64 ÷ 8	8	(79) 6 ÷ 1	6
(5) 3 ÷ 1	3	(30) 56 ÷ 8	7	(55) 6 ÷ 2	3	(80) 36 ÷ 6	6
(6) 42 ÷ 6	7	(31) 25 ÷ 5	5	(56) 70 ÷ 7	10	(81) 28 ÷ 7	4
(7) 6 ÷ 3	2	(32) 1 ÷ 1	1	(57) 12 ÷ 4	3	(82) 24 ÷ 4	6
(8) 10 ÷ 1	10	(33) 14 ÷ 7	2	(58) 4 ÷ 1	4	(83) 3 ÷ 1	3
(9) 24 ÷ 3	8	(34) 30 ÷ 3	10	(59) 30 ÷ 6	5	(84) 24 ÷ 3	8
(10) 20 ÷ 5	4	(35) 7 ÷ 1	7	(60) 40 ÷ 8	5	(85) 63 ÷ 7	9
(11) 6 ÷ 6	1	(36) 16 ÷ 4	4	(61) 7 ÷ 7	1	(86) 32 ÷ 8	4
(12) 63 ÷ 7	9	(37) 72 ÷ 8	9	(62) 10 ÷ 2	5	(87) 56 ÷ 7	8
(13) 48 ÷ 8	6	(38) 10 ÷ 5	2	(63) 49 ÷ 7	7	(88) 40 ÷ 5	8
(14) 8 ÷ 4	2	(39) 9 ÷ 1	9	(64) 20 ÷ 4	5	(89) 42 ÷ 7	6
(15) 32 ÷ 8	4	(40) 28 ÷ 4	7	(65) 2 ÷ 1	2	(90) 25 ÷ 5	5
(16) 18 ÷ 3	6	(41) 12 ÷ 6	2	(66) 27 ÷ 3	9	(91) 24 ÷ 6	4
(17) 16 ÷ 2	8	(42) 15 ÷ 3	5	(67) 21 ÷ 7	3	(92) 14 ÷ 7	2
(18) 56 ÷ 7	8	(43) 8 ÷ 1	8	(68) 45 ÷ 5	9	(93) 28 ÷ 4	7
(19) 24 ÷ 8	3	(44) 35 ÷ 5	7	(69) 20 ÷ 2	10	(94) 64 ÷ 8	8
(20) 40 ÷ 5	8	(45) 18 ÷ 6	3	(70) 4 ÷ 4	1	(95) 35 ÷ 5	7
(21) 5 ÷ 1	5	(46) 12 ÷ 3	4	(71) 35 ÷ 7	5	(96) 40 ÷ 8	5
(22) 2 ÷ 2	1	(47) 40 ÷ 4	10	(72) 48 ÷ 6	8	(97) 72 ÷ 8	9
(23) 42 ÷ 7	6	(48) 16 ÷ 8	2	(73) 21 ÷ 3	7	(98) 49 ÷ 7	7
(24) 80 ÷ 8	10	(49) 3 ÷ 3	1	(74) 15 ÷ 5	3	(99) 64 ÷ 8	8
(25) 24 ÷ 6	4	(50) 18 ÷ 2	9	(75) 8 ÷ 8	1	(100) 18 ÷ 6	3

Division 9

Name_____ Date _____ Score _____ Time _____

(1) $54 \div 9$	(26) $45 \div 9$	(51) $90 \div 9$	(76) $63 \div 9$
(2) $81 \div 9$	(27) $36 \div 9$	(52) $18 \div 9$	(77) $54 \div 9$
(3) $27 \div 9$	(28) $18 \div 9$	(53) $63 \div 9$	(78) $9 \div 9$
(4) $45 \div 9$	(29) $90 \div 9$	(54) $27 \div 9$	(79) $81 \div 9$
(5) $72 \div 9$	(30) $9 \div 9$	(55) $72 \div 9$	(80) $36 \div 9$
(6) $18 \div 9$	(31) $72 \div 9$	(56) $45 \div 9$	(81) $18 \div 9$
(7) $63 \div 9$	(32) $54 \div 9$	(57) $9 \div 9$	(82) $72 \div 9$
(8) $90 \div 9$	(33) $45 \div 9$	(58) $36 \div 9$	(83) $45 \div 9$
(9) $36 \div 9$	(34) $27 \div 9$	(59) $18 \div 9$	(84) $9 \div 9$
(10) $9 \div 9$	(35) $81 \div 9$	(60) $72 \div 9$	(85) $27 \div 9$
(11) $72 \div 9$	(36) $63 \div 9$	(61) $54 \div 9$	(86) $36 \div 9$
(12) $27 \div 9$	(37) $18 \div 9$	(62) $81 \div 9$	(87) $63 \div 9$
(13) $45 \div 9$	(38) $36 \div 9$	(63) $63 \div 9$	(88) $18 \div 9$
(14) $63 \div 9$	(39) $9 \div 9$	(64) $45 \div 9$	(89) $90 \div 9$
(15) $81 \div 9$	(40) $54 \div 9$	(65) $27 \div 9$	(90) $36 \div 9$
(16) $18 \div 9$	(41) $72 \div 9$	(66) $54 \div 9$	(91) $81 \div 9$
(17) $54 \div 9$	(42) $27 \div 9$	(67) $90 \div 9$	(92) $9 \div 9$
(18) $36 \div 9$	(43) $63 \div 9$	(68) $36 \div 9$	(93) $72 \div 9$
(19) $27 \div 9$	(44) $36 \div 9$	(69) $9 \div 9$	(94) $54 \div 9$
(20) $90 \div 9$	(45) $18 \div 9$	(70) $72 \div 9$	(95) $45 \div 9$
(21) $63 \div 9$	(46) $45 \div 9$	(71) $81 \div 9$	(96) $27 \div 9$
(22) $72 \div 9$	(47) $90 \div 9$	(72) $54 \div 9$	(97) $63 \div 9$
(23) $9 \div 9$	(48) $81 \div 9$	(73) $63 \div 9$	(98) $90 \div 9$
(24) $81 \div 9$	(49) $54 \div 9$	(74) $27 \div 9$	(99) $36 \div 9$
(25) $27 \div 9$	(50) $9 \div 9$	(75) $90 \div 9$	(100) $45 \div 9$

34

Division 9

(1) 54 ÷ 9	6	(26) 45 ÷ 9	5	(51) 90 ÷ 9	10	(76) 63 ÷ 9	7
(2) 81 ÷ 9	9	(27) 36 ÷ 9	4	(52) 18 ÷ 9	2	(77) 54 ÷ 9	6
(3) 27 ÷ 9	3	(28) 18 ÷ 9	2	(53) 63 ÷ 9	7	(78) 9 ÷ 9	1
(4) 45 ÷ 9	5	(29) 90 ÷ 9	10	(54) 27 ÷ 9	3	(79) 81 ÷ 9	9
(5) 72 ÷ 9	8	(30) 9 ÷ 9	1	(55) 72 ÷ 9	8	(80) 36 ÷ 9	4
(6) 18 ÷ 9	2	(31) 72 ÷ 9	8	(56) 45 ÷ 9	5	(81) 18 ÷ 9	2
(7) 63 ÷ 9	7	(32) 54 ÷ 9	6	(57) 9 ÷ 9	1	(82) 72 ÷ 9	8
(8) 90 ÷ 9	10	(33) 45 ÷ 9	5	(58) 36 ÷ 9	4	(83) 45 ÷ 9	5
(9) 36 ÷ 9	4	(34) 27 ÷ 9	3	(59) 18 ÷ 9	2	(84) 9 ÷ 9	1
(10) 9 ÷ 9	1	(35) 81 ÷ 9	9	(60) 72 ÷ 9	8	(85) 27 ÷ 9	3
(11) 72 ÷ 9	8	(36) 63 ÷ 9	7	(61) 54 ÷ 9	6	(86) 36 ÷ 9	4
(12) 27 ÷ 9	3	(37) 18 ÷ 9	2	(62) 81 ÷ 9	9	(87) 63 ÷ 9	7
(13) 45 ÷ 9	5	(38) 36 ÷ 9	4	(63) 63 ÷ 9	7	(88) 18 ÷ 9	2
(14) 63 ÷ 9	7	(39) 9 ÷ 9	1	(64) 45 ÷ 9	5	(89) 90 ÷ 9	10
(15) 81 ÷ 9	9	(40) 54 ÷ 9	6	(65) 27 ÷ 9	3	(90) 36 ÷ 9	4
(16) 18 ÷ 9	2	(41) 72 ÷ 9	8	(66) 54 ÷ 9	6	(91) 81 ÷ 9	9
(17) 54 ÷ 9	6	(42) 27 ÷ 9	3	(67) 90 ÷ 9	10	(92) 9 ÷ 9	1
(18) 36 ÷ 9	4	(43) 63 ÷ 9	7	(68) 36 ÷ 9	4	(93) 72 ÷ 9	8
(19) 27 ÷ 9	3	(44) 36 ÷ 9	4	(69) 9 ÷ 9	1	(94) 54 ÷ 9	6
(20) 90 ÷ 9	10	(45) 18 ÷ 9	2	(70) 72 ÷ 9	8	(95) 45 ÷ 9	5
(21) 63 ÷ 9	7	(46) 45 ÷ 9	5	(71) 81 ÷ 9	9	(96) 27 ÷ 9	3
(22) 72 ÷ 9	8	(47) 90 ÷ 9	10	(72) 54 ÷ 9	6	(97) 63 ÷ 9	7
(23) 9 ÷ 9	1	(48) 81 ÷ 9	9	(73) 63 ÷ 9	7	(98) 90 ÷ 9	10
(24) 81 ÷ 9	9	(49) 54 ÷ 9	6	(74) 27 ÷ 9	3	(99) 36 ÷ 9	4
(25) 27 ÷ 9	3	(50) 9 ÷ 9	1	(75) 90 ÷ 9	10	(100) 45 ÷ 9	5

Division 1-9

Name_____ Date_____ Score _____ Time _____

(1) $9 \div 3$		(26) $63 \div 9$		(51) $54 \div 6$		(76) $72 \div 8$	
(2) $35 \div 5$		(27) $3 \div 3$		(52) $21 \div 3$		(77) $4 \div 4$	
(3) $45 \div 9$		(28) $25 \div 5$		(53) $8 \div 4$		(78) $30 \div 3$	
(4) $49 \div 7$		(29) $18 \div 2$		(54) $27 \div 9$		(79) $28 \div 7$	
(5) $32 \div 4$		(30) $4 \div 1$		(55) $70 \div 7$		(80) $16 \div 2$	
(6) $10 \div 2$		(31) $35 \div 7$		(56) $12 \div 3$		(81) $5 \div 1$	
(7) $2 \div 1$		(32) $40 \div 5$		(57) $14 \div 2$		(82) $15 \div 5$	
(8) $48 \div 6$		(33) $12 \div 2$		(58) $6 \div 6$		(83) $72 \div 9$	
(9) $32 \div 8$		(34) $30 \div 6$		(59) $56 \div 8$		(84) $2 \div 2$	
(10) $81 \div 9$		(35) $20 \div 4$		(60) $36 \div 4$		(85) $24 \div 3$	
(11) $50 \div 5$		(36) $90 \div 9$		(61) $20 \div 2$		(86) $3 \div 1$	
(12) $12 \div 4$		(37) $6 \div 2$		(62) $18 \div 9$		(87) $60 \div 6$	
(13) $7 \div 1$		(38) $16 \div 8$		(63) $42 \div 6$		(88) $8 \div 8$	
(14) $36 \div 9$		(39) $40 \div 4$		(64) $6 \div 3$		(89) $36 \div 6$	
(15) $18 \div 6$		(40) $56 \div 7$		(65) $8 \div 1$		(90) $24 \div 6$	
(16) $27 \div 3$		(41) $15 \div 3$		(66) $30 \div 5$		(91) $35 \div 5$	
(17) $10 \div 1$		(42) $6 \div 1$		(67) $24 \div 8$		(92) $10 \div 2$	
(18) $64 \div 8$		(43) $10 \div 5$		(68) $42 \div 7$		(93) $48 \div 6$	
(19) $14 \div 7$		(44) $54 \div 9$		(69) $8 \div 2$		(94) $81 \div 9$	
(20) $63 \div 7$		(45) $80 \div 8$		(70) $28 \div 4$		(95) $36 \div 9$	
(21) $18 \div 3$		(46) $16 \div 4$		(71) $9 \div 1$		(96) $64 \div 8$	
(22) $4 \div 2$		(47) $1 \div 1$		(72) $24 \div 4$		(97) $42 \div 6$	
(23) $20 \div 5$		(48) $12 \div 6$		(73) $5 \div 5$		(98) $63 \div 7$	
(24) $48 \div 8$		(49) $7 \div 7$		(74) $40 \div 8$		(99) $45 \div 9$	
(25) $9 \div 9$		(50) $45 \div 5$		(75) $21 \div 7$		(100) $27 \div 9$	

Division 1-9

(1) $9 \div 3$	3	
(2) $35 \div 5$	7	
(3) $45 \div 9$	5	
(4) $49 \div 7$	7	
(5) $32 \div 4$	8	
(6) $10 \div 2$	5	
(7) $2 \div 1$	2	
(8) $48 \div 6$	8	
(9) $32 \div 8$	4	
(10) $81 \div 9$	9	
(11) $50 \div 5$	10	
(12) $12 \div 4$	3	
(13) $7 \div 1$	7	
(14) $36 \div 9$	4	
(15) $18 \div 6$	3	
(16) $27 \div 3$	9	
(17) $10 \div 1$	10	
(18) $64 \div 8$	8	
(19) $14 \div 7$	2	
(20) $63 \div 7$	9	
(21) $18 \div 3$	6	
(22) $4 \div 2$	2	
(23) $20 \div 5$	4	
(24) $48 \div 8$	6	
(25) $9 \div 9$	1	

(26) $63 \div 9$	7	
(27) $3 \div 3$	1	
(28) $25 \div 5$	5	
(29) $18 \div 2$	9	
(30) $4 \div 1$	4	
(31) $35 \div 7$	5	
(32) $40 \div 5$	8	
(33) $12 \div 2$	6	
(34) $30 \div 6$	5	
(35) $20 \div 4$	5	
(36) $90 \div 9$	10	
(37) $6 \div 2$	3	
(38) $16 \div 8$	2	
(39) $40 \div 4$	10	
(40) $56 \div 7$	8	
(41) $15 \div 3$	5	
(42) $6 \div 1$	6	
(43) $10 \div 5$	2	
(44) $54 \div 9$	6	
(45) $80 \div 8$	10	
(46) $16 \div 4$	4	
(47) $1 \div 1$	1	
(48) $12 \div 6$	2	
(49) $7 \div 7$	1	
(50) $45 \div 5$	9	

(51) $54 \div 6$	9	
(52) $21 \div 3$	7	
(53) $8 \div 4$	2	
(54) $27 \div 9$	3	
(55) $70 \div 7$	10	
(56) $12 \div 3$	4	
(57) $14 \div 2$	7	
(58) $6 \div 6$	1	
(59) $56 \div 8$	7	
(60) $36 \div 4$	9	
(61) $20 \div 2$	10	
(62) $18 \div 9$	2	
(63) $42 \div 6$	7	
(64) $6 \div 3$	2	
(65) $8 \div 1$	8	
(66) $30 \div 5$	6	
(67) $24 \div 8$	3	
(68) $42 \div 7$	6	
(69) $8 \div 2$	4	
(70) $28 \div 4$	7	
(71) $9 \div 1$	9	
(72) $24 \div 4$	6	
(73) $5 \div 5$	1	
(74) $40 \div 8$	5	
(75) $21 \div 7$	3	

(76) $72 \div 8$	9	
(77) $4 \div 4$	1	
(78) $30 \div 3$	10	
(79) $28 \div 7$	4	
(80) $16 \div 2$	8	
(81) $5 \div 1$	5	
(82) $15 \div 5$	3	
(83) $72 \div 9$	8	
(84) $2 \div 2$	1	
(85) $24 \div 3$	8	
(86) $3 \div 1$	3	
(87) $60 \div 6$	10	
(88) $8 \div 8$	1	
(89) $36 \div 6$	6	
(90) $24 \div 6$	4	
(91) $35 \div 5$	7	
(92) $10 \div 2$	5	
(93) $48 \div 6$	8	
(94) $81 \div 9$	9	
(95) $36 \div 9$	4	
(96) $64 \div 8$	8	
(97) $42 \div 6$	7	
(98) $63 \div 7$	9	
(99) $45 \div 9$	5	
(100) $27 \div 9$	3	

Division 10

Name_____ Date_____ Score_____ Time_____

(1) $40 \div 10$		(26) $90 \div 10$		(51) $20 \div 10$		(76) $100 \div 10$	
(2) $100 \div 10$		(27) $50 \div 10$		(52) $70 \div 10$		(77) $50 \div 10$	
(3) $20 \div 10$		(28) $100 \div 10$		(53) $50 \div 10$		(78) $10 \div 10$	
(4) $90 \div 10$		(29) $30 \div 10$		(54) $60 \div 10$		(79) $60 \div 10$	
(5) $10 \div 10$		(30) $40 \div 10$		(55) $80 \div 10$		(80) $40 \div 10$	
(6) $80 \div 10$		(31) $70 \div 10$		(56) $30 \div 10$		(81) $90 \div 10$	
(7) $50 \div 10$		(32) $20 \div 10$		(57) $10 \div 10$		(82) $70 \div 10$	
(8) $70 \div 10$		(33) $50 \div 10$		(58) $100 \div 10$		(83) $30 \div 10$	
(9) $30 \div 10$		(34) $60 \div 10$		(59) $90 \div 10$		(84) $80 \div 10$	
(10) $90 \div 10$		(35) $10 \div 10$		(60) $70 \div 10$		(85) $10 \div 10$	
(11) $60 \div 10$		(36) $80 \div 10$		(61) $40 \div 10$		(86) $20 \div 10$	
(12) $20 \div 10$		(37) $40 \div 10$		(62) $30 \div 10$		(87) $100 \div 10$	
(13) $80 \div 10$		(38) $90 \div 10$		(63) $50 \div 10$		(88) $50 \div 10$	
(14) $40 \div 10$		(39) $20 \div 10$		(64) $10 \div 10$		(89) $80 \div 10$	
(15) $100 \div 10$		(40) $50 \div 10$		(65) $80 \div 10$		(90) $40 \div 10$	
(16) $10 \div 10$		(41) $70 \div 10$		(66) $60 \div 10$		(91) $30 \div 10$	
(17) $70 \div 10$		(42) $60 \div 10$		(67) $20 \div 10$		(92) $90 \div 10$	
(18) $50 \div 10$		(43) $30 \div 10$		(68) $100 \div 10$		(93) $10 \div 10$	
(19) $80 \div 10$		(44) $100 \div 10$		(69) $50 \div 10$		(94) $70 \div 10$	
(20) $20 \div 10$		(45) $80 \div 10$		(70) $40 \div 10$		(95) $50 \div 10$	
(21) $60 \div 10$		(46) $10 \div 10$		(71) $90 \div 10$		(96) $60 \div 10$	
(22) $90 \div 10$		(47) $70 \div 10$		(72) $30 \div 10$		(97) $80 \div 10$	
(23) $30 \div 10$		(48) $40 \div 10$		(73) $20 \div 10$		(98) $20 \div 10$	
(24) $100 \div 10$		(49) $60 \div 10$		(74) $70 \div 10$		(99) $30 \div 10$	
(25) $40 \div 10$		(50) $50 \div 10$		(75) $10 \div 10$		(100) $100 \div 10$	

Division 10

(1) 40 ÷ 10	4	(26) 90 ÷ 10	9	(51) 20 ÷ 10	2	(76) 100 ÷ 10	10
(2) 100 ÷ 10	10	(27) 50 ÷ 10	5	(52) 70 ÷ 10	7	(77) 50 ÷ 10	5
(3) 20 ÷ 10	2	(28) 100 ÷ 10	10	(53) 50 ÷ 10	5	(78) 10 ÷ 10	1
(4) 90 ÷ 10	9	(29) 30 ÷ 10	3	(54) 60 ÷ 10	6	(79) 60 ÷ 10	6
(5) 10 ÷ 10	1	(30) 40 ÷ 10	4	(55) 80 ÷ 10	8	(80) 40 ÷ 10	4
(6) 80 ÷ 10	8	(31) 70 ÷ 10	7	(56) 30 ÷ 10	3	(81) 90 ÷ 10	9
(7) 50 ÷ 10	5	(32) 20 ÷ 10	2	(57) 10 ÷ 10	1	(82) 70 ÷ 10	7
(8) 70 ÷ 10	7	(33) 50 ÷ 10	5	(58) 100 ÷ 10	10	(83) 30 ÷ 10	3
(9) 30 ÷ 10	3	(34) 60 ÷ 10	6	(59) 90 ÷ 10	9	(84) 80 ÷ 10	8
(10) 90 ÷ 10	9	(35) 10 ÷ 10	1	(60) 70 ÷ 10	7	(85) 10 ÷ 10	1
(11) 60 ÷ 10	6	(36) 80 ÷ 10	8	(61) 40 ÷ 10	4	(86) 20 ÷ 10	2
(12) 20 ÷ 10	2	(37) 40 ÷ 10	4	(62) 30 ÷ 10	3	(87) 100 ÷ 10	10
(13) 80 ÷ 10	8	(38) 90 ÷ 10	9	(63) 50 ÷ 10	5	(88) 50 ÷ 10	5
(14) 40 ÷ 10	4	(39) 20 ÷ 10	2	(64) 10 ÷ 10	1	(89) 80 ÷ 10	8
(15) 100 ÷ 10	10	(40) 50 ÷ 10	5	(65) 80 ÷ 10	8	(90) 40 ÷ 10	4
(16) 10 ÷ 10	1	(41) 70 ÷ 10	7	(66) 60 ÷ 10	6	(91) 30 ÷ 10	3
(17) 70 ÷ 10	7	(42) 60 ÷ 10	6	(67) 20 ÷ 10	2	(92) 90 ÷ 10	9
(18) 50 ÷ 10	5	(43) 30 ÷ 10	3	(68) 100 ÷ 10	10	(93) 10 ÷ 10	1
(19) 80 ÷ 10	8	(44) 100 ÷ 10	10	(69) 50 ÷ 10	5	(94) 70 ÷ 10	7
(20) 20 ÷ 10	2	(45) 80 ÷ 10	8	(70) 40 ÷ 10	4	(95) 50 ÷ 10	5
(21) 60 ÷ 10	6	(46) 10 ÷ 10	1	(71) 90 ÷ 10	9	(96) 60 ÷ 10	6
(22) 90 ÷ 10	9	(47) 70 ÷ 10	7	(72) 30 ÷ 10	3	(97) 80 ÷ 10	8
(23) 30 ÷ 10	3	(48) 40 ÷ 10	4	(73) 20 ÷ 10	2	(98) 20 ÷ 10	2
(24) 100 ÷ 10	10	(49) 60 ÷ 10	6	(74) 70 ÷ 10	7	(99) 30 ÷ 10	3
(25) 40 ÷ 10	4	(50) 50 ÷ 10	5	(75) 10 ÷ 10	1	(100) 100 ÷ 10	10

Division I

Name_____ Date_____ Score_____ Time_____

(1) $15 \div 5$		(26) $20 \div 4$		(51) $18 \div 2$		(76) $100 \div 10$	
(2) $27 \div 9$		(27) $30 \div 5$		(52) $56 \div 7$		(77) $12 \div 2$	
(3) $72 \div 9$		(28) $72 \div 8$		(53) $20 \div 10$		(78) $27 \div 9$	
(4) $7 \div 1$		(29) $28 \div 7$		(54) $10 \div 5$		(79) $60 \div 6$	
(5) $54 \div 6$		(30) $21 \div 3$		(55) $80 \div 8$		(80) $24 \div 8$	
(6) $63 \div 9$		(31) $24 \div 4$		(56) $4 \div 4$		(81) $40 \div 8$	
(7) $30 \div 3$		(32) $12 \div 3$		(57) $32 \div 8$		(82) $8 \div 4$	
(8) $8 \div 4$		(33) $36 \div 6$		(58) $30 \div 10$		(83) $30 \div 10$	
(9) $40 \div 8$		(34) $2 \div 1$		(59) $6 \div 6$		(84) $63 \div 9$	
(10) $24 \div 3$		(35) $4 \div 2$		(60) $3 \div 1$		(85) $54 \div 6$	
(11) $60 \div 10$		(36) $40 \div 4$		(61) $40 \div 5$		(86) $7 \div 1$	
(12) $54 \div 9$		(37) $16 \div 8$		(62) $6 \div 3$		(87) $72 \div 8$	
(13) $36 \div 4$		(38) $56 \div 8$		(63) $60 \div 6$		(88) $20 \div 5$	
(14) $81 \div 9$		(39) $14 \div 2$		(64) $36 \div 9$		(89) $16 \div 2$	
(15) $28 \div 7$		(40) $42 \div 7$		(65) $18 \div 9$		(90) $12 \div 6$	
(16) $16 \div 2$		(41) $9 \div 3$		(66) $48 \div 6$		(91) $27 \div 3$	
(17) $10 \div 1$		(42) $20 \div 2$		(67) $12 \div 4$		(92) $36 \div 6$	
(18) $70 \div 7$		(43) $21 \div 7$		(68) $18 \div 3$		(93) $50 \div 10$	
(19) $1 \div 1$		(44) $90 \div 10$		(69) $64 \div 8$		(94) $72 \div 8$	
(20) $42 \div 6$		(45) $14 \div 7$		(70) $32 \div 4$		(95) $56 \div 7$	
(21) $35 \div 5$		(46) $30 \div 6$		(71) $6 \div 2$		(96) $4 \div 1$	
(22) $45 \div 9$		(47) $15 \div 3$		(72) $40 \div 10$		(97) $12 \div 6$	
(23) $80 \div 10$		(48) $18 \div 6$		(73) $63 \div 7$		(98) $21 \div 7$	
(24) $25 \div 5$		(49) $10 \div 1$		(74) $45 \div 5$		(99) $35 \div 5$	
(25) $10 \div 2$		(50) $24 \div 3$		(75) $16 \div 4$		(100) $14 \div 7$	

Division I

40

(1) $15 \div 5$	3	(26) $20 \div 4$	5	(51) $18 \div 2$	9	(76) $100 \div 10$	10
(2) $27 \div 9$	3	(27) $30 \div 5$	6	(52) $56 \div 7$	8	(77) $12 \div 2$	6
(3) $72 \div 9$	8	(28) $72 \div 8$	9	(53) $20 \div 10$	2	(78) $27 \div 9$	3
(4) $7 \div 1$	7	(29) $28 \div 7$	4	(54) $10 \div 5$	2	(79) $60 \div 6$	10
(5) $54 \div 6$	9	(30) $21 \div 3$	7	(55) $80 \div 8$	10	(80) $24 \div 8$	3
(6) $63 \div 9$	7	(31) $24 \div 4$	6	(56) $4 \div 4$	1	(81) $40 \div 8$	5
(7) $30 \div 3$	10	(32) $12 \div 3$	4	(57) $32 \div 8$	4	(82) $8 \div 4$	2
(8) $8 \div 4$	2	(33) $36 \div 6$	6	(58) $30 \div 10$	3	(83) $30 \div 10$	3
(9) $40 \div 8$	5	(34) $2 \div 1$	2	(59) $6 \div 6$	1	(84) $63 \div 9$	7
(10) $24 \div 3$	8	(35) $4 \div 2$	2	(60) $3 \div 1$	3	(85) $54 \div 6$	9
(11) $60 \div 10$	6	(36) $40 \div 4$	10	(61) $40 \div 5$	8	(86) $7 \div 1$	7
(12) $54 \div 9$	6	(37) $16 \div 8$	2	(62) $6 \div 3$	2	(87) $72 \div 8$	9
(13) $36 \div 4$	9	(38) $56 \div 8$	7	(63) $60 \div 6$	10	(88) $20 \div 5$	4
(14) $81 \div 9$	9	(39) $14 \div 2$	7	(64) $36 \div 9$	4	(89) $16 \div 2$	8
(15) $28 \div 7$	4	(40) $42 \div 7$	6	(65) $18 \div 9$	2	(90) $12 \div 6$	2
(16) $16 \div 2$	8	(41) $9 \div 3$	3	(66) $48 \div 6$	8	(91) $27 \div 3$	9
(17) $10 \div 1$	10	(42) $20 \div 2$	10	(67) $12 \div 4$	3	(92) $36 \div 6$	6
(18) $70 \div 7$	10	(43) $21 \div 7$	3	(68) $18 \div 3$	6	(93) $50 \div 10$	5
(19) $1 \div 1$	1	(44) $90 \div 10$	9	(69) $64 \div 8$	8	(94) $72 \div 8$	9
(20) $42 \div 6$	7	(45) $14 \div 7$	2	(70) $32 \div 4$	8	(95) $56 \div 7$	8
(21) $35 \div 5$	7	(46) $30 \div 6$	5	(71) $6 \div 2$	3	(96) $4 \div 1$	4
(22) $45 \div 9$	5	(47) $15 \div 3$	5	(72) $40 \div 10$	4	(97) $12 \div 6$	2
(23) $80 \div 10$	8	(48) $18 \div 6$	3	(73) $63 \div 7$	9	(98) $21 \div 7$	3
(24) $25 \div 5$	5	(49) $10 \div 1$	10	(74) $45 \div 5$	9	(99) $35 \div 5$	7
(25) $10 \div 2$	5	(50) $24 \div 3$	8	(75) $16 \div 4$	4	(100) $14 \div 7$	2

Division II

Name_____ Date_____ Score_____ Time_____

(1) $14 \div 7$		(26) $30 \div 5$		(51) $15 \div 5$		(76) $16 \div 4$	
(2) $21 \div 7$		(27) $28 \div 7$		(52) $72 \div 9$		(77) $45 \div 5$	
(3) $4 \div 1$		(28) $24 \div 4$		(53) $54 \div 6$		(78) $63 \div 7$	
(4) $72 \div 8$		(29) $36 \div 6$		(54) $30 \div 3$		(79) $40 \div 10$	
(5) $36 \div 6$		(30) $4 \div 2$		(55) $40 \div 8$		(80) $6 \div 2$	
(6) $12 \div 6$		(31) $16 \div 8$		(56) $60 \div 10$		(81) $32 \div 4$	
(7) $20 \div 5$		(32) $14 \div 2$		(57) $36 \div 4$		(82) $64 \div 8$	
(8) $7 \div 1$		(33) $9 \div 3$		(58) $28 \div 7$		(83) $18 \div 3$	
(9) $63 \div 9$		(34) $21 \div 7$		(59) $10 \div 1$		(84) $12 \div 4$	
(10) $8 \div 4$		(35) $14 \div 7$		(60) $1 \div 1$		(85) $48 \div 6$	
(11) $24 \div 8$		(36) $15 \div 3$		(61) $35 \div 5$		(86) $18 \div 9$	
(12) $27 \div 9$		(37) $10 \div 1$		(62) $80 \div 10$		(87) $36 \div 9$	
(13) $100 \div 10$		(38) $24 \div 3$		(63) $10 \div 2$		(88) $60 \div 6$	
(14) $35 \div 5$		(39) $18 \div 6$		(64) $25 \div 5$		(89) $6 \div 3$	
(15) $12 \div 6$		(40) $30 \div 6$		(65) $45 \div 9$		(90) $40 \div 5$	
(16) $56 \div 7$		(41) $90 \div 10$		(66) $42 \div 6$		(91) $3 \div 1$	
(17) $50 \div 10$		(42) $20 \div 2$		(67) $70 \div 7$		(92) $6 \div 6$	
(18) $27 \div 3$		(43) $42 \div 7$		(68) $16 \div 2$		(93) $30 \div 10$	
(19) $16 \div 2$		(44) $56 \div 8$		(69) $81 \div 9$		(94) $32 \div 8$	
(20) $72 \div 9$		(45) $40 \div 4$		(70) $54 \div 9$		(95) $4 \div 4$	
(21) $54 \div 6$		(46) $4 \div 2$		(71) $24 \div 3$		(96) $80 \div 8$	
(22) $30 \div 10$		(47) $36 \div 9$		(72) $8 \div 4$		(97) $10 \div 5$	
(23) $40 \div 8$		(48) $2 \div 1$		(73) $63 \div 9$		(98) $20 \div 10$	
(24) $60 \div 6$		(49) $12 \div 3$		(74) $7 \div 1$		(99) $56 \div 7$	
(25) $12 \div 2$		(50) $21 \div 3$		(75) $27 \div 9$		(100) $18 \div 2$	

Division II

(1) $14 \div 7$	2	(26) $30 \div 5$	6	(51) $15 \div 5$	3	(76) $16 \div 4$	4
(2) $21 \div 7$	3	(27) $28 \div 7$	4	(52) $72 \div 9$	8	(77) $45 \div 5$	9
(3) $4 \div 1$	4	(28) $24 \div 4$	6	(53) $54 \div 6$	9	(78) $63 \div 7$	9
(4) $72 \div 8$	9	(29) $36 \div 6$	6	(54) $30 \div 3$	10	(79) $40 \div 10$	4
(5) $36 \div 6$	6	(30) $4 \div 2$	2	(55) $40 \div 8$	5	(80) $6 \div 2$	3
(6) $12 \div 6$	2	(31) $16 \div 8$	2	(56) $60 \div 10$	6	(81) $32 \div 4$	8
(7) $20 \div 5$	4	(32) $14 \div 2$	7	(57) $36 \div 4$	9	(82) $64 \div 8$	8
(8) $7 \div 1$	7	(33) $9 \div 3$	3	(58) $28 \div 7$	4	(83) $18 \div 3$	6
(9) $63 \div 9$	7	(34) $21 \div 7$	3	(59) $10 \div 1$	10	(84) $12 \div 4$	3
(10) $8 \div 4$	2	(35) $14 \div 7$	2	(60) $1 \div 1$	1	(85) $48 \div 6$	8
(11) $24 \div 8$	3	(36) $15 \div 3$	5	(61) $35 \div 5$	7	(86) $18 \div 9$	2
(12) $27 \div 9$	3	(37) $10 \div 1$	10	(62) $80 \div 10$	8	(87) $36 \div 9$	4
(13) $100 \div 10$	10	(38) $24 \div 3$	8	(63) $10 \div 2$	5	(88) $60 \div 6$	10
(14) $35 \div 5$	7	(39) $18 \div 6$	3	(64) $25 \div 5$	5	(89) $6 \div 3$	2
(15) $12 \div 6$	2	(40) $30 \div 6$	5	(65) $45 \div 9$	5	(90) $40 \div 5$	8
(16) $56 \div 7$	8	(41) $90 \div 10$	9	(66) $42 \div 6$	7	(91) $3 \div 1$	3
(17) $50 \div 10$	5	(42) $20 \div 2$	10	(67) $70 \div 7$	10	(92) $6 \div 6$	1
(18) $27 \div 3$	9	(43) $42 \div 7$	6	(68) $16 \div 2$	8	(93) $30 \div 10$	3
(19) $16 \div 2$	8	(44) $56 \div 8$	7	(69) $81 \div 9$	9	(94) $32 \div 8$	4
(20) $72 \div 9$	8	(45) $40 \div 4$	10	(70) $54 \div 9$	6	(95) $4 \div 4$	1
(21) $54 \div 6$	9	(46) $4 \div 2$	2	(71) $24 \div 3$	8	(96) $80 \div 8$	10
(22) $30 \div 10$	3	(47) $36 \div 9$	4	(72) $8 \div 4$	2	(97) $10 \div 5$	2
(23) $40 \div 8$	5	(48) $2 \div 1$	2	(73) $63 \div 9$	7	(98) $20 \div 10$	2
(24) $60 \div 6$	10	(49) $12 \div 3$	4	(74) $7 \div 1$	7	(99) $56 \div 7$	8
(25) $12 \div 2$	6	(50) $21 \div 3$	7	(75) $27 \div 9$	3	(100) $18 \div 2$	9

Division III

Name_____ Date _____ Score _____ Time _____

(1) $63 \div 7$		(26) $9 \div 3$		(51) $50 \div 10$		(76) $90 \div 10$	
(2) $36 \div 9$		(27) $25 \div 5$		(52) $14 \div 7$		(77) $20 \div 4$	
(3) $40 \div 8$		(28) $16 \div 2$		(53) $48 \div 6$		(78) $4 \div 2$	
(4) $8 \div 2$		(29) $81 \div 9$		(54) $6 \div 2$		(79) $21 \div 7$	
(5) $12 \div 6$		(30) $30 \div 6$		(55) $32 \div 4$		(80) $6 \div 6$	
(6) $9 \div 1$		(31) $49 \div 7$		(56) $10 \div 1$		(81) $70 \div 7$	
(7) $12 \div 4$		(32) $18 \div 9$		(57) $70 \div 10$		(82) $28 \div 4$	
(8) $60 \div 10$		(33) $3 \div 3$		(58) $15 \div 5$		(83) $30 \div 5$	
(9) $18 \div 3$		(34) $35 \div 5$		(59) $6 \div 3$		(84) $10 \div 2$	
(10) $64 \div 8$		(35) $24 \div 3$		(60) $56 \div 8$		(85) $3 \div 1$	
(11) $7 \div 7$		(36) $4 \div 1$		(61) $27 \div 9$		(86) $40 \div 10$	
(12) $2 \div 1$		(37) $8 \div 8$		(62) $7 \div 1$		(87) $12 \div 3$	
(13) $24 \div 4$		(38) $30 \div 10$		(63) $24 \div 6$		(88) $32 \div 8$	
(14) $100 \div 10$		(39) $15 \div 3$		(64) $72 \div 8$		(89) $90 \div 9$	
(15) $40 \div 8$		(40) $12 \div 2$		(65) $40 \div 4$		(90) $21 \div 3$	
(16) $42 \div 6$		(41) $80 \div 8$		(66) $35 \div 7$		(91) $56 \div 7$	
(17) $63 \div 9$		(42) $42 \div 7$		(67) $27 \div 3$		(92) $20 \div 5$	
(18) $10 \div 5$		(43) $2 \div 2$		(68) $20 \div 2$		(93) $9 \div 9$	
(19) $30 \div 3$		(44) $8 \div 4$		(69) $1 \div 1$		(94) $54 \div 9$	
(20) $14 \div 2$		(45) $36 \div 6$		(70) $45 \div 5$		(95) $50 \div 5$	
(21) $6 \div 1$		(46) $60 \div 6$		(71) $24 \div 8$		(96) $4 \div 4$	
(22) $16 \div 8$		(47) $16 \div 4$		(72) $80 \div 10$		(97) $72 \div 9$	
(23) $10 \div 10$		(48) $5 \div 5$		(73) $48 \div 8$		(98) $5 \div 1$	
(24) $36 \div 4$		(49) $8 \div 1$		(74) $18 \div 6$		(99) $20 \div 10$	
(25) $28 \div 7$		(50) $45 \div 9$		(75) $21 \div 3$		(100) $18 \div 2$	

Division III

(1) 63 ÷ 7	9	
(2) 36 ÷ 9	4	
(3) 40 ÷ 8	5	
(4) 8 ÷ 2	4	
(5) 12 ÷ 6	2	
(6) 9 ÷ 1	9	
(7) 12 ÷ 4	3	
(8) 60 ÷ 10	6	
(9) 18 ÷ 3	6	
(10) 64 ÷ 8	8	
(11) 7 ÷ 7	1	
(12) 2 ÷ 1	2	
(13) 24 ÷ 4	6	
(14) 100 ÷ 10	10	
(15) 40 ÷ 8	5	
(16) 42 ÷ 6	7	
(17) 63 ÷ 9	7	
(18) 10 ÷ 5	2	
(19) 30 ÷ 3	10	
(20) 14 ÷ 2	7	
(21) 6 ÷ 1	6	
(22) 16 ÷ 8	2	
(23) 10 ÷ 10	1	
(24) 36 ÷ 4	9	
(25) 28 ÷ 7	4	

(26) 9 ÷ 3	3	
(27) 25 ÷ 5	5	
(28) 16 ÷ 2	8	
(29) 81 ÷ 9	9	
(30) 30 ÷ 6	5	
(31) 49 ÷ 7	7	
(32) 18 ÷ 9	2	
(33) 3 ÷ 3	1	
(34) 35 ÷ 5	7	
(35) 24 ÷ 3	8	
(36) 4 ÷ 1	4	
(37) 8 ÷ 8	1	
(38) 30 ÷ 10	3	
(39) 15 ÷ 3	5	
(40) 12 ÷ 2	6	
(41) 80 ÷ 8	10	
(42) 42 ÷ 7	6	
(43) 2 ÷ 2	1	
(44) 8 ÷ 4	2	
(45) 36 ÷ 6	6	
(46) 60 ÷ 6	10	
(47) 16 ÷ 4	4	
(48) 5 ÷ 5	1	
(49) 8 ÷ 1	8	
(50) 45 ÷ 9	5	

(51) 50 ÷ 10	5	
(52) 14 ÷ 7	2	
(53) 48 ÷ 6	8	
(54) 6 ÷ 2	3	
(55) 32 ÷ 4	8	
(56) 10 ÷ 1	10	
(57) 70 ÷ 10	7	
(58) 15 ÷ 5	3	
(59) 6 ÷ 3	2	
(60) 56 ÷ 8	7	
(61) 27 ÷ 9	3	
(62) 7 ÷ 1	7	
(63) 24 ÷ 6	4	
(64) 72 ÷ 8	9	
(65) 40 ÷ 4	10	
(66) 35 ÷ 7	5	
(67) 27 ÷ 3	9	
(68) 20 ÷ 2	10	
(69) 1 ÷ 1	1	
(70) 45 ÷ 5	9	
(71) 24 ÷ 8	3	
(72) 80 ÷ 10	8	
(73) 48 ÷ 8	6	
(74) 18 ÷ 6	3	
(75) 21 ÷ 3	7	

(76) 90 ÷ 10	9	
(77) 20 ÷ 4	5	
(78) 4 ÷ 2	2	
(79) 21 ÷ 7	3	
(80) 6 ÷ 6	1	
(81) 70 ÷ 7	10	
(82) 28 ÷ 4	7	
(83) 30 ÷ 5	6	
(84) 10 ÷ 2	5	
(85) 3 ÷ 1	3	
(86) 40 ÷ 10	4	
(87) 12 ÷ 3	4	
(88) 32 ÷ 8	4	
(89) 90 ÷ 9	10	
(90) 21 ÷ 3	7	
(91) 56 ÷ 7	8	
(92) 20 ÷ 5	4	
(93) 9 ÷ 9	1	
(94) 54 ÷ 9	6	
(95) 50 ÷ 5	10	
(96) 4 ÷ 4	1	
(97) 72 ÷ 9	8	
(98) 5 ÷ 1	5	
(99) 20 ÷ 10	2	
(100) 18 ÷ 2	9	

Division IV

Name_____ Date _____ Score _____ Time _____

(1) $21 \div 3$		(26) $28 \div 7$		(51) $18 \div 2$		(76) $45 \div 9$	
(2) $18 \div 6$		(27) $36 \div 4$		(52) $20 \div 10$		(77) $8 \div 1$	
(3) $48 \div 8$		(28) $10 \div 10$		(53) $5 \div 1$		(78) $5 \div 5$	
(4) $80 \div 10$		(29) $16 \div 8$		(54) $72 \div 9$		(79) $16 \div 4$	
(5) $24 \div 8$		(30) $6 \div 1$		(55) $4 \div 4$		(80) $60 \div 6$	
(6) $45 \div 5$		(31) $14 \div 2$		(56) $50 \div 5$		(81) $36 \div 6$	
(7) $1 \div 1$		(32) $30 \div 3$		(57) $54 \div 9$		(82) $8 \div 4$	
(8) $20 \div 2$		(33) $10 \div 5$		(58) $9 \div 9$		(83) $2 \div 2$	
(9) $27 \div 3$		(34) $63 \div 9$		(59) $20 \div 5$		(84) $42 \div 7$	
(10) $35 \div 7$		(35) $42 \div 6$		(60) $56 \div 7$		(85) $80 \div 8$	
(11) $40 \div 4$		(36) $40 \div 8$		(61) $21 \div 3$		(86) $12 \div 2$	
(12) $72 \div 8$		(37) $100 \div 10$		(62) $32 \div 8$		(87) $15 \div 3$	
(13) $24 \div 6$		(38) $24 \div 4$		(63) $90 \div 9$		(88) $30 \div 10$	
(14) $7 \div 1$		(39) $7 \div 7$		(64) $12 \div 3$		(89) $24 \div 3$	
(15) $27 \div 9$		(40) $2 \div 1$		(65) $40 \div 10$		(90) $8 \div 8$	
(16) $56 \div 8$		(41) $64 \div 8$		(66) $3 \div 1$		(91) $35 \div 5$	
(17) $6 \div 3$		(42) $18 \div 3$		(67) $10 \div 2$		(92) $4 \div 1$	
(18) $15 \div 5$		(43) $60 \div 10$		(68) $30 \div 5$		(93) $3 \div 3$	
(19) $70 \div 10$		(44) $12 \div 4$		(69) $28 \div 4$		(94) $18 \div 9$	
(20) $10 \div 1$		(45) $9 \div 1$		(70) $70 \div 7$		(95) $49 \div 7$	
(21) $32 \div 4$		(46) $12 \div 6$		(71) $6 \div 6$		(96) $30 \div 6$	
(22) $6 \div 2$		(47) $40 \div 8$		(72) $21 \div 7$		(97) $81 \div 9$	
(23) $48 \div 6$		(48) $8 \div 2$		(73) $20 \div 4$		(98) $16 \div 2$	
(24) $14 \div 7$		(49) $36 \div 9$		(74) $4 \div 2$		(99) $25 \div 5$	
(25) $50 \div 10$		(50) $63 \div 7$		(75) $90 \div 10$		(100) $9 \div 3$	

Division IV

(1) $21 \div 3$	7	(26) $28 \div 7$	4	(51) $18 \div 2$	9	(76) $45 \div 9$	5
(2) $18 \div 6$	3	(27) $36 \div 4$	9	(52) $20 \div 10$	2	(77) $8 \div 1$	8
(3) $48 \div 8$	6	(28) $10 \div 10$	1	(53) $5 \div 1$	5	(78) $5 \div 5$	1
(4) $80 \div 10$	8	(29) $16 \div 8$	2	(54) $72 \div 9$	8	(79) $16 \div 4$	4
(5) $24 \div 8$	3	(30) $6 \div 1$	6	(55) $4 \div 4$	1	(80) $60 \div 6$	10
(6) $45 \div 5$	9	(31) $14 \div 2$	7	(56) $50 \div 5$	10	(81) $36 \div 6$	6
(7) $1 \div 1$	1	(32) $30 \div 3$	10	(57) $54 \div 9$	6	(82) $8 \div 4$	2
(8) $20 \div 2$	10	(33) $10 \div 5$	2	(58) $9 \div 9$	1	(83) $2 \div 2$	1
(9) $27 \div 3$	9	(34) $63 \div 9$	7	(59) $20 \div 5$	4	(84) $42 \div 7$	6
(10) $35 \div 7$	5	(35) $42 \div 6$	7	(60) $56 \div 7$	8	(85) $80 \div 8$	10
(11) $40 \div 4$	10	(36) $40 \div 8$	5	(61) $21 \div 3$	7	(86) $12 \div 2$	6
(12) $72 \div 8$	9	(37) $100 \div 10$	10	(62) $32 \div 8$	4	(87) $15 \div 3$	5
(13) $24 \div 6$	4	(38) $24 \div 4$	6	(63) $90 \div 9$	10	(88) $30 \div 10$	3
(14) $7 \div 1$	7	(39) $7 \div 7$	1	(64) $12 \div 3$	4	(89) $24 \div 3$	8
(15) $27 \div 9$	3	(40) $2 \div 1$	2	(65) $40 \div 10$	4	(90) $8 \div 8$	1
(16) $56 \div 8$	7	(41) $64 \div 8$	8	(66) $3 \div 1$	3	(91) $35 \div 5$	7
(17) $6 \div 3$	2	(42) $18 \div 3$	6	(67) $10 \div 2$	5	(92) $4 \div 1$	4
(18) $15 \div 5$	3	(43) $60 \div 10$	6	(68) $30 \div 5$	6	(93) $3 \div 3$	1
(19) $70 \div 10$	7	(44) $12 \div 4$	3	(69) $28 \div 4$	7	(94) $18 \div 9$	2
(20) $10 \div 1$	10	(45) $9 \div 1$	9	(70) $70 \div 7$	10	(95) $49 \div 7$	7
(21) $32 \div 4$	8	(46) $12 \div 6$	2	(71) $6 \div 6$	1	(96) $30 \div 6$	5
(22) $6 \div 2$	3	(47) $40 \div 8$	5	(72) $21 \div 7$	3	(97) $81 \div 9$	9
(23) $48 \div 6$	8	(48) $8 \div 2$	4	(73) $20 \div 4$	5	(98) $16 \div 2$	8
(24) $14 \div 7$	2	(49) $36 \div 9$	4	(74) $4 \div 2$	2	(99) $25 \div 5$	5
(25) $50 \div 10$	5	(50) $63 \div 7$	9	(75) $90 \div 10$	9	(100) $9 \div 3$	3

Division V

Name_____ Date _____ Score _____ Time _____

(1) $56 \div 8$		(26) $70 \div 7$		(51) $3 \div 3$		(76) $60 \div 10$	
(2) $15 \div 3$		(27) $18 \div 2$		(52) $18 \div 9$		(77) $9 \div 3$	
(3) $4 \div 1$		(28) $72 \div 9$		(53) $12 \div 4$		(78) $2 \div 2$	
(4) $12 \div 6$		(29) $21 \div 3$		(54) $4 \div 4$		(79) $14 \div 2$	
(5) $54 \div 9$		(30) $5 \div 1$		(55) $60 \div 6$		(80) $3 \div 1$	
(6) $20 \div 4$		(31) $50 \div 5$		(56) $10 \div 5$		(81) $90 \div 10$	
(7) $42 \div 6$		(32) $4 \div 2$		(57) $64 \div 8$		(82) $30 \div 6$	
(8) $45 \div 5$		(33) $48 \div 6$		(58) $25 \div 5$		(83) $8 \div 2$	
(9) $28 \div 4$		(34) $32 \div 4$		(59) $100 \div 10$		(84) $40 \div 8$	
(10) $63 \div 7$		(35) $10 \div 1$		(60) $48 \div 8$		(85) $7 \div 1$	
(11) $16 \div 4$		(36) $15 \div 5$		(61) $9 \div 9$		(86) $27 \div 3$	
(12) $49 \div 7$		(37) $6 \div 3$		(62) $80 \div 8$		(87) $63 \div 9$	
(13) $20 \div 2$		(38) $70 \div 10$		(63) $28 \div 7$		(88) $10 \div 10$	
(14) $6 \div 2$		(39) $24 \div 6$		(64) $24 \div 3$		(89) $35 \div 7$	
(15) $40 \div 4$		(40) $24 \div 8$		(65) $40 \div 8$		(90) $2 \div 1$	
(16) $36 \div 6$		(41) $18 \div 6$		(66) $9 \div 1$		(91) $72 \div 8$	
(17) $56 \div 7$		(42) $45 \div 9$		(67) $27 \div 9$		(92) $21 \div 3$	
(18) $20 \div 10$		(43) $5 \div 5$		(68) $40 \div 10$		(93) $30 \div 3$	
(19) $12 \div 3$		(44) $8 \div 4$		(69) $30 \div 5$		(94) $16 \div 2$	
(20) $81 \div 9$		(45) $42 \div 7$		(70) $12 \div 2$		(95) $6 \div 1$	
(21) $14 \div 7$		(46) $8 \div 8$		(71) $7 \div 7$		(96) $80 \div 10$	
(22) $24 \div 4$		(47) $10 \div 2$		(72) $18 \div 3$		(97) $30 \div 10$	
(23) $1 \div 1$		(48) $35 \div 5$		(73) $90 \div 9$		(98) $36 \div 4$	
(24) $32 \div 8$		(49) $36 \div 9$		(74) $6 \div 6$		(99) $50 \div 10$	
(25) $20 \div 5$		(50) $16 \div 8$		(75) $8 \div 1$		(100) $21 \div 7$	

Division V

(1) 56 ÷ 8	7	(26) 70 ÷ 7	10	(51) 3 ÷ 3	1	(76) 60 ÷ 10	6
(2) 15 ÷ 3	5	(27) 18 ÷ 2	9	(52) 18 ÷ 9	2	(77) 9 ÷ 3	3
(3) 4 ÷ 1	4	(28) 72 ÷ 9	8	(53) 12 ÷ 4	3	(78) 2 ÷ 2	1
(4) 12 ÷ 6	2	(29) 21 ÷ 3	7	(54) 4 ÷ 4	1	(79) 14 ÷ 2	7
(5) 54 ÷ 9	6	(30) 5 ÷ 1	5	(55) 60 ÷ 6	10	(80) 3 ÷ 1	3
(6) 20 ÷ 4	5	(31) 50 ÷ 5	10	(56) 10 ÷ 5	2	(81) 90 ÷ 10	9
(7) 42 ÷ 6	7	(32) 4 ÷ 2	2	(57) 64 ÷ 8	8	(82) 30 ÷ 6	5
(8) 45 ÷ 5	9	(33) 48 ÷ 6	8	(58) 25 ÷ 5	5	(83) 8 ÷ 2	4
(9) 28 ÷ 4	7	(34) 32 ÷ 4	8	(59) 100 ÷ 10	10	(84) 40 ÷ 8	5
(10) 63 ÷ 7	9	(35) 10 ÷ 1	10	(60) 48 ÷ 8	6	(85) 7 ÷ 1	7
(11) 16 ÷ 4	4	(36) 15 ÷ 5	3	(61) 9 ÷ 9	1	(86) 27 ÷ 3	9
(12) 49 ÷ 7	7	(37) 6 ÷ 3	2	(62) 80 ÷ 8	10	(87) 63 ÷ 9	7
(13) 20 ÷ 2	10	(38) 70 ÷ 10	7	(63) 28 ÷ 7	4	(88) 10 ÷ 10	1
(14) 6 ÷ 2	3	(39) 24 ÷ 6	4	(64) 24 ÷ 3	8	(89) 35 ÷ 7	5
(15) 40 ÷ 4	10	(40) 24 ÷ 8	3	(65) 40 ÷ 8	5	(90) 2 ÷ 1	2
(16) 36 ÷ 6	6	(41) 18 ÷ 6	3	(66) 9 ÷ 1	9	(91) 72 ÷ 8	9
(17) 56 ÷ 7	8	(42) 45 ÷ 9	5	(67) 27 ÷ 9	3	(92) 21 ÷ 3	7
(18) 20 ÷ 10	2	(43) 5 ÷ 5	1	(68) 40 ÷ 10	4	(93) 30 ÷ 3	10
(19) 12 ÷ 3	4	(44) 8 ÷ 4	2	(69) 30 ÷ 5	6	(94) 16 ÷ 2	8
(20) 81 ÷ 9	9	(45) 42 ÷ 7	6	(70) 12 ÷ 2	6	(95) 6 ÷ 1	6
(21) 14 ÷ 7	2	(46) 8 ÷ 8	1	(71) 7 ÷ 7	1	(96) 80 ÷ 10	8
(22) 24 ÷ 4	6	(47) 10 ÷ 2	5	(72) 18 ÷ 3	6	(97) 30 ÷ 10	3
(23) 1 ÷ 1	1	(48) 35 ÷ 5	7	(73) 90 ÷ 9	10	(98) 36 ÷ 4	9
(24) 32 ÷ 8	4	(49) 36 ÷ 9	4	(74) 6 ÷ 6	1	(99) 50 ÷ 10	5
(25) 20 ÷ 5	4	(50) 16 ÷ 8	2	(75) 8 ÷ 1	8	(100) 21 ÷ 7	3

FOLDERS
AND
RECORDS

FOLDERS AND RECORDS

This section contains reproducible pages and forms to help you and your students keep track of progress toward mastering division facts.

Student Folders

Each student in my class has a folder. It can be a plain manila folder or a pocket portfolio. My students have pocket portfolios color-coded for each major subject. Staple or tape any (or all) of the following into the folder. Tailor it to your own preferences!

Division Stickers, p. 52. I staple this inside the folder on the left. Tests which are completed in three minutes or less are graded by the teacher. If a child gets 100% correct, note the date the test was passed on this sheet and award a sticker.

Record of Timed Tests, p. 53. I duplicate four of these sheets, one for each quarter, and note the days and dates school will be in session (no weekends or holidays) under "Date." For example: T 9/5/01, W 9/6/01, Th 9/7/01, F 9/8/01, etc. I use these as masters and copy a set for each student on colored paper. They are stapled into the folders on the right. After each daily test, each child notes the test she took: 1-2, 3, 1-3, etc. or a ditto symbol if the same test was taken as on the previous day. The child then notes her score and time. If a child is absent, the note **Absent** can be written in. After results are recorded, the test papers can be recycled or discarded.

Graph, p. 54. For review tests on addition, subtraction, and multiplication, I allow students only *three* minutes (rather than five) to complete the 100 facts. They make a line graph of their *scores* as they progress toward completing 100 in three minutes. The only students who note their times during review are those who have already passed and continue trying to improve their speeds.

Multiples Chart, p. 51. A chart of multiples can be attached to the folder. Teach students how to use it to find answers. This relieves pressure on students, especially those for whom learning the facts is difficult, and especially when they're first beginning with a new timed test. Alternately, you may wish to post a large chart on the wall (in the back of the room so it's not *too* easy to see). Because of the repetition built into the tests, students won't need to refer to the chart for long. Checking a chart takes time. It is nearly impossible to pass in three minutes or less if it's used.

Answer Keys, pp. 8-48. Once students begin progressing at their own speeds through the timed test worksheets, they will need answer keys so they can check their own papers on days when they do not finish in less than three minutes. I duplicate all division keys for each child two-sided on yellow paper and staple them together to be kept in the folders. I collect them for reuse at the end of the school year.

Flash Cards, pp. 59-78. Flash cards in plastic sandwich bags may also be stored in the folders until they're sent home.

Teacher Record-Keeping

Time Chart. A large chart like the example below can be posted in the front of the room. The teacher (or an aide) points to each time as it elapses after a timed test begins. When a child completes the timed test, he/she looks up and writes the time to which the teacher is pointing.

1:00	2:00	3:00	4:00
1:05	2:05	3:05	4:05
1:10	2:10	3:10	4:10
1:15	2:15	3:15	4:15
1:20	2:20	3:20	4:20
1:25	2:25	3:25	4:25
1:30	2:30	3:30	4:30
1:35	2:35	3:35	4:35
1:40	2:40	3:40	4:40
1:45	2:45	3:45	4:45
1:50	2:50	3:50	4:50
1:55	2:55	3:55	4:55
			5:00

Class Record, p. 57. The chart can be duplicated or a gradebook can be used. List student names on the left. When a child passes a test, check the box or make a small note of the date the test was passed. An alternative is to use a large wall chart. Children may write in the dates they pass.

Congratulations Certificates, p. 58. In place of or in addition to the sticker in the folder, you can fill out a certificate for a child to take home and decorate *it* with a sticker.

MULTIPLES

The chart of multiples on the right may be reproduced and glued into the folders behind the Record of Timed Tests, where it is available but not *easily* available. Teach students how to use it for division facts. See p. 50 for alternatives.

1	2	3	4	5	6	7	8	9	10
2	4	6	8	10	12	14	16	18	20
3	6	9	12	15	18	21	24	27	30
4	8	12	16	20	24	28	32	36	40
5	10	15	20	25	30	35	40	45	50
6	12	18	24	30	36	42	48	54	60
7	14	21	28	35	42	49	56	63	70
8	16	24	32	40	48	56	64	72	80
9	18	27	36	45	54	63	72	81	90
10	20	30	40	50	60	70	80	90	100

Division Stickers

1-2	Date passed _____
3	Date passed _____
1-3	Date passed _____
4	Date passed _____
1-4	Date passed _____
5	Date passed _____
1-5	Date passed _____
6	Date passed _____
1-6	Date passed _____

7	Date passed _____
1-7	Date passed _____
8	Date passed _____
1-8	Date passed _____
9	Date passed _____
1-9	Date passed _____
10	Date passed _____
1-10	Date passed _____

Record of Timed Tests

Name_____

DATE	TEST	SCORE	TIME	DATE	TEST	SCORE	TIME

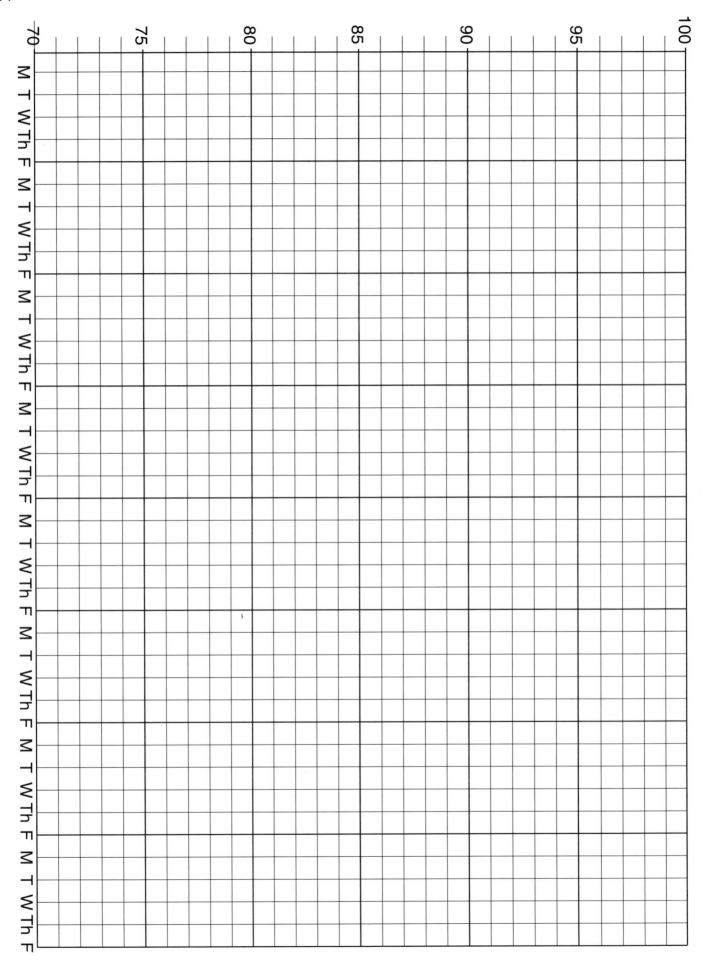

Division	1-2	3	1-3	4	1-4	5	1-5	6	1-6	7	1-7	8	1-8	9	1-9	10	ALL

CONGRATULATIONS!

_____ completed

100 division facts correctly in three

minutes or less today and passed the

_____ test! This is an accomplishment

to *celebrate!*

Date _____

Teacher _____

CONGRATULATIONS!

_____ completed

100 division facts correctly in three

minutes or less today and passed the

_____ test! This is an accomplishment

to *celebrate!*

Date _____

Teacher _____

CONGRATULATIONS!

_____ completed

100 division facts correctly in three

minutes or less today and passed the

_____ test! This is an accomplishment

to *celebrate!*

Date _____

Teacher _____

CONGRATULATIONS!

_____ completed

100 division facts correctly in three

minutes or less today and passed the

_____ test! This is an accomplishment

to *celebrate!*

Date _____

Teacher _____

FLASH
CARD
MASTERS

Flash Cards

On the following pages are black-line masters for personal flash cards. Here are some ideas and possibilities for using them:

1) Flash cards can be run two-sided on tagboard or heavy paper.

2) Students can practice a set of facts with flash cards alone or with a friend before attempting the timed tests.

3) Students can store personal flash cards in plastic sandwich bags in their math folders or in boxes in their desks.

4) Flash cards can be run one-sided and used for games such as Concentration. Students can create games for each other to play.

5) Flash cards can be sent home with students for practice with parents.

6) Addition, subtraction, multiplication and division flash cards can be color-coded.

$$3 \div 1 =$$

$$6 \div 1 =$$

$$9 \div 1 =$$

ONES

$$2 \div 1 =$$

$$5 \div 1 =$$

$$8 \div 1 =$$

$$1 \div 1 =$$

$$4 \div 1 =$$

$$7 \div 1 =$$

$$10 \div 1 =$$

60

$1 \div 1 = 1$
$2 \div 1 = 2$
$3 \div 1 = 3$
$4 \div 1 = 4$
$5 \div 1 = 5$
$6 \div 1 = 6$
$7 \div 1 = 7$
$8 \div 1 = 8$
$9 \div 1 = 9$
$10 \div 1 = 10$

$3 \div 1 = \boxed{3}$

$6 \div 1 = \boxed{6}$

$9 \div 1 = \boxed{9}$

$2 \div 1 = \boxed{2}$

$5 \div 1 = \boxed{5}$

$8 \div 1 = \boxed{8}$

$1 \div 1 = \boxed{1}$

$4 \div 1 = \boxed{4}$

$7 \div 1 = \boxed{7}$

$10 \div 1 = \boxed{10}$

$6 \div 2 =$	$12 \div 2 =$	$18 \div 2 =$
$4 \div 2 =$	$10 \div 2 =$	$16 \div 2 =$
$2 \div 2 =$	$8 \div 2 =$	$14 \div 2 =$

TWOS

$20 \div 2 =$

62

2 ÷ 2 = 1
4 ÷ 2 = 2
6 ÷ 2 = 3
8 ÷ 2 = 4
10 ÷ 2 = 5
12 ÷ 2 = 6
14 ÷ 2 = 7
16 ÷ 2 = 8
18 ÷ 2 = 9
20 ÷ 2 = 10

6 ÷ 2 = 3

4 ÷ 2 = 2

2 ÷ 2 = 1

12 ÷ 2 = 6

10 ÷ 2 = 5

8 ÷ 2 = 4

18 ÷ 2 = 9

16 ÷ 2 = 8

14 ÷ 2 = 7

20 ÷ 2 = 10

$9 \div 3 =$

$6 \div 3 =$

$3 \div 3 =$

$18 \div 3 =$

$15 \div 3 =$

$12 \div 3 =$

$27 \div 3 =$

$24 \div 3 =$

$21 \div 3 =$

THREES

$30 \div 3 =$

64

$3 \div 3 = 1$
$6 \div 3 = 2$
$9 \div 3 = 3$
$12 \div 3 = 4$
$15 \div 3 = 5$
$18 \div 3 = 6$
$21 \div 3 = 7$
$24 \div 3 = 8$
$27 \div 3 = 9$
$30 \div 3 = 10$

$9 \div 3 = \boxed{3}$

$18 \div 3 = \boxed{6}$

$27 \div 3 = \boxed{9}$

$6 \div 3 = \boxed{2}$

$15 \div 3 = \boxed{5}$

$24 \div 3 = \boxed{8}$

$3 \div 3 = \boxed{1}$

$12 \div 3 = \boxed{14}$

$21 \div 3 = \boxed{27}$

$30 \div 3 = \boxed{10}$

$12 \div 4 =$

$24 \div 4 =$

$36 \div 4 =$

FOURS

$8 \div 4 =$

$20 \div 4 =$

$32 \div 4 =$

$4 \div 4 =$

$16 \div 4 =$

$28 \div 4 =$

$40 \div 4 =$

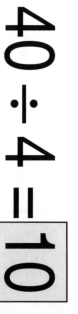

$4 \div 4 = 1$
$8 \div 4 = 2$
$12 \div 4 = 3$
$16 \div 4 = 4$
$20 \div 4 = 5$
$24 \div 4 = 6$
$28 \div 4 = 7$
$32 \div 4 = 8$
$36 \div 4 = 9$
$40 \div 4 = 10$

$12 \div 4 = 3$

$24 \div 4 = 6$

$36 \div 4 = 9$

$8 \div 4 = 2$

$20 \div 4 = 5$

$32 \div 4 = 8$

$4 \div 4 = 1$

$16 \div 4 = 24$

$28 \div 4 = 37$

$40 \div 4 = 10$

$5 \div 5 =$	$10 \div 5 =$	$15 \div 5 =$
$20 \div 5 =$	$25 \div 5 =$	$30 \div 5 =$
$35 \div 5 =$	$40 \div 5 =$	$45 \div 5 =$
$50 \div 5 =$		*FIVES*

68

5 ÷ 5 =   1
10 ÷ 5 =   2
15 ÷ 5 =   3
20 ÷ 5 =   4
25 ÷ 5 =   5
30 ÷ 5 =   6
35 ÷ 5 =   7
40 ÷ 5 =   8
45 ÷ 5 =   9
50 ÷ 5 = 10

$$15 \div 5 = \boxed{3}$$

$$30 \div 5 = \boxed{6}$$

$$45 \div 5 = \boxed{9}$$

$$10 \div 5 = \boxed{2}$$

$$25 \div 5 = \boxed{5}$$

$$40 \div 5 = \boxed{8}$$

$$5 \div 5 = \boxed{1}$$

$$20 \div 5 = \boxed{4}$$

$$35 \div 5 = \boxed{7}$$

$$50 \div 5 = \boxed{10}$$

18 ÷ 6 =

36 ÷ 6 =

54 ÷ 6 =

SIXES

12 ÷ 6 =

30 ÷ 6 =

48 ÷ 6 =

6 ÷ 6 =

24 ÷ 6 =

42 ÷ 6 =

60 ÷ 6 =

70

$6 \div 6 = 1$
$12 \div 6 = 2$
$18 \div 6 = 3$
$24 \div 6 = 4$
$30 \div 6 = 5$
$36 \div 6 = 6$
$42 \div 6 = 7$
$48 \div 6 = 8$
$54 \div 6 = 9$
$60 \div 6 = 10$

$18 \div 6 = \boxed{3}$

$12 \div 6 = \boxed{2}$

$6 \div 6 = \boxed{1}$

$36 \div 6 = \boxed{6}$

$30 \div 6 = \boxed{5}$

$24 \div 6 = \boxed{4}$

$54 \div 6 = \boxed{9}$

$48 \div 6 = \boxed{8}$

$42 \div 6 = \boxed{7}$

$60 \div 6 = \boxed{10}$

$21 \div 7 =$

$42 \div 7 =$

$63 \div 7 =$

SEVENS

$14 \div 7 =$

$35 \div 7 =$

$56 \div 7 =$

$7 \div 7 =$

$28 \div 7 =$

$49 \div 7 =$

$70 \div 7 =$

72

$7 \div 7 = \quad 1$
$14 \div 7 = \quad 2$
$21 \div 7 = \quad 3$
$28 \div 7 = \quad 4$
$35 \div 7 = \quad 5$
$42 \div 7 = \quad 6$
$49 \div 7 = \quad 7$
$56 \div 7 = \quad 8$
$63 \div 7 = \quad 9$
$70 \div 7 = 10$

$21 \div 7 = \boxed{3}$

$42 \div 7 = \boxed{6}$

$63 \div 7 = \boxed{9}$

$14 \div 7 = \boxed{2}$

$35 \div 7 = \boxed{5}$

$56 \div 7 = \boxed{8}$

$7 \div 7 = \boxed{21}$

$28 \div 7 = \boxed{44}$

$49 \div 7 = \boxed{67}$

$70 \div 7 = \boxed{10}$

24 ÷ 8 =

48 ÷ 8 =

72 ÷ 8 =

EIGHTS

16 ÷ 8 =

40 ÷ 8 =

64 ÷ 8 =

8 ÷ 8 =

32 ÷ 8 =

56 ÷ 8 =

80 ÷ 8 =

74

$8 \div 8 = 1$
$16 \div 8 = 2$
$24 \div 8 = 3$
$32 \div 8 = 4$
$40 \div 8 = 5$
$48 \div 8 = 6$
$56 \div 8 = 7$
$64 \div 8 = 8$
$72 \div 8 = 9$
$80 \div 8 = 10$

$24 \div 8 = \boxed{3}$

$48 \div 8 = \boxed{6}$

$72 \div 8 = \boxed{9}$

$16 \div 8 = \boxed{2}$

$40 \div 8 = \boxed{5}$

$64 \div 8 = \boxed{8}$

$8 \div 8 = \boxed{21}$

$32 \div 8 = \boxed{4}$

$56 \div 8 = \boxed{7}$

$80 \div 8 = \boxed{10}$

$27 \div 9 =$

$54 \div 9 =$

$81 \div 9 =$

NINES

$18 \div 9 =$

$45 \div 9 =$

$72 \div 9 =$

$9 \div 9 =$

$36 \div 9 =$

$63 \div 9 =$

$90 \div 9 =$

76

$9 \div 9 = 1$
$18 \div 9 = 2$
$27 \div 9 = 3$
$36 \div 9 = 4$
$45 \div 9 = 5$
$54 \div 9 = 6$
$63 \div 9 = 7$
$72 \div 9 = 8$
$81 \div 9 = 9$
$90 \div 9 = 10$

$27 \div 9 = \boxed{3}$

$54 \div 9 = \boxed{6}$

$81 \div 9 = \boxed{9}$

$18 \div 9 = \boxed{2}$

$45 \div 9 = \boxed{5}$

$72 \div 9 = \boxed{8}$

$9 \div 9 = \boxed{21}$

$36 \div 9 = \boxed{54}$

$63 \div 9 = \boxed{7}$

$90 \div 9 = \boxed{10}$

30 ÷ 10 =

60 ÷ 10 =

90 ÷ 10 =

TENS

20 ÷ 10 =

50 ÷ 10 =

80 ÷ 10 =

10 ÷ 10 =

40 ÷ 10 =

70 ÷ 10 =

100 ÷ 10 =

78

$10 \div 10 = 1$
$20 \div 10 = 2$
$30 \div 10 = 3$
$40 \div 10 = 4$
$50 \div 10 = 5$
$60 \div 10 = 6$
$70 \div 10 = 7$
$80 \div 10 = 8$
$90 \div 10 = 9$
$100 \div 10 = 10$

$30 \div 10 = \boxed{3}$

$60 \div 10 = \boxed{6}$

$90 \div 10 = \boxed{9}$

$20 \div 10 = \boxed{2}$

$50 \div 10 = \boxed{5}$

$80 \div 10 = \boxed{8}$

$10 \div 10 = \boxed{1}$

$40 \div 10 = \boxed{4}$

$70 \div 10 = \boxed{7}$

$100 \div 10 = \boxed{10}$

REVIEW
MASTERS

REVIEW MASTERS

Early in the school year, I review addition, subtraction, and multiplication facts to bring students "up to speed" and to give students who have not yet memorized these facts a chance to learn them. For this, I generally allow only *three* minutes for a test, rather than five. This is a group activity, with all students doing the same worksheet at once. After time has been called, I read the answers aloud as students check their work, and they keep track of their scores with the graph on p. 56. Students who finish in three minutes or less with 100% accuracy receive certificates. They may then continue to do review tests to improve their speeds.

When all or most students have demonstrated mastery of the addition facts, I repeat the process with subtraction facts, again working toward the goal of all or most students completing 100 facts in three minutes or less with 100% accuracy. Unless students have practiced the facts in earlier years, it may be more difficult for them to learn addition and subtraction facts than multiplication and division facts.

Consider finishing the review with a challenge to the principal, other teachers, or older students. There is a chance that one of the fastest students in your class will complete the worksheet in less time than an adult or older student who is out of practice!

Addition

Name_____ Date _____ Score _____ Time _____

(1) 1 + 1		(26) 10 + 6		(51) 7 + 6		(76) 7 + 10	
(2) 4 + 2		(27) 10 + 10		(52) 10 + 2		(77) 10 + 1	
(3) 9 + 6		(28) 6 + 9		(53) 2 + 7		(78) 8 + 6	
(4) 1 + 4		(29) 10 + 8		(54) 8 + 2		(79) 7 + 7	
(5) 9 + 10		(30) 3 + 10		(55) 2 + 4		(80) 4 + 8	
(6) 5 + 4		(31) 3 + 1		(56) 10 + 5		(81) 0 + 2	
(7) 6 + 10		(32) 5 + 10		(57) 9 + 7		(82) 4 + 6	
(8) 2 + 10		(33) 0 + 1		(58) 0 + 4		(83) 8 + 7	
(9) 5 + 7		(34) 9 + 8		(59) 9 + 3		(84) 8 + 10	
(10) 4 + 9		(35) 3 + 3		(60) 7 + 5		(85) 7 + 3	
(11) 4 + 4		(36) 7 + 8		(61) 9 + 5		(86) 9 + 9	
(12) 5 + 5		(37) 6 + 5		(62) 3 + 4		(87) 3 + 8	
(13) 10 + 4		(38) 1 + 9		(63) 1 + 3		(88) 1 + 5	
(14) 8 + 4		(39) 10 + 3		(64) 8 + 8		(89) 6 + 2	
(15) 4 + 5		(40) 1 + 2		(65) 0 + 10		(90) 2 + 1	
(16) 5 + 8		(41) 8 + 1		(66) 3 + 7		(91) 7 + 1	
(17) 9 + 4		(42) 5 + 3		(67) 7 + 2		(92) 6 + 7	
(18) 6 + 5		(43) 1 + 8		(68) 5 + 2		(93) 1 + 6	
(19) 0 + 8		(44) 5 + 9		(69) 10 + 9		(94) 2 + 9	
(20) 3 + 6		(45) 4 + 1		(70) 4 + 7		(95) 2 + 5	
(21) 3 + 5		(46) 6 + 8		(71) 8 + 5		(96) 3 + 4	
(22) 10 + 7		(47) 6 + 1		(72) 2 + 6		(97) 8 + 3	
(23) 7 + 9		(48) 9 + 2		(73) 3 + 9		(98) 2 + 3	
(24) 4 + 10		(49) 6 + 3		(74) 6 + 6		(99) 2 + 8	
(25) 6 + 4		(50) 7 + 4		(75) 8 + 9		(100) 2 + 2	

82

Addition

(1)	1 + 1	2	(26)	10 + 6	16	(51)	7 + 6	13	(76)	7 + 10	17
(2)	4 + 2	6	(27)	10 + 10	20	(52)	10 + 2	12	(77)	10 + 1	11
(3)	9 + 6	15	(28)	6 + 9	15	(53)	2 + 7	9	(78)	8 + 6	14
(4)	1 + 4	5	(29)	10 + 8	18	(54)	8 + 2	10	(79)	7 + 7	14
(5)	9 + 10	19	(30)	3 + 10	13	(55)	2 + 4	6	(80)	4 + 8	12
(6)	5 + 4	9	(31)	3 + 1	4	(56)	10 + 5	15	(81)	0 + 2	2
(7)	6 + 10	16	(32)	5 + 10	15	(57)	9 + 7	16	(82)	4 + 6	10
(8)	2 + 10	12	(33)	0 + 1	1	(58)	0 + 4	4	(83)	8 + 7	15
(9)	5 + 7	12	(34)	9 + 8	17	(59)	9 + 3	12	(84)	8 + 10	18
(10)	4 + 9	13	(35)	3 + 3	6	(60)	7 + 5	12	(85)	7 + 3	10
(11)	4 + 4	8	(36)	7 + 8	15	(61)	9 + 5	14	(86)	9 + 9	18
(12)	5 + 5	10	(37)	6 + 5	11	(62)	3 + 4	7	(87)	3 + 8	11
(13)	10 + 4	14	(38)	1 + 9	10	(63)	1 + 3	4	(88)	1 + 5	6
(14)	8 + 4	12	(39)	10 + 3	13	(64)	8 + 8	16	(89)	6 + 2	8
(15)	4 + 5	9	(40)	1 + 2	3	(65)	0 + 10	10	(90)	2 + 1	3
(16)	5 + 8	13	(41)	8 + 1	9	(66)	3 + 7	10	(91)	7 + 1	8
(17)	9 + 4	13	(42)	5 + 3	8	(67)	7 + 2	9	(92)	6 + 7	13
(18)	6 + 5	11	(43)	1 + 8	9	(68)	5 + 2	7	(93)	1 + 6	7
(19)	0 + 8	8	(44)	5 + 9	14	(69)	10 + 9	19	(94)	2 + 9	11
(20)	3 + 6	9	(45)	4 + 1	5	(70)	4 + 7	11	(95)	2 + 5	7
(21)	3 + 5	8	(46)	6 + 8	14	(71)	8 + 5	13	(96)	3 + 4	7
(22)	10 + 7	17	(47)	6 + 1	7	(72)	2 + 6	8	(97)	8 + 3	11
(23)	7 + 9	16	(48)	9 + 2	11	(73)	3 + 9	12	(98)	2 + 3	5
(24)	4 + 10	14	(49)	6 + 3	9	(74)	6 + 6	12	(99)	2 + 8	10
(25)	6 + 4	10	(50)	7 + 4	11	(75)	8 + 9	17	(100)	2 + 2	4

83

Subtraction

Name_____ Date_____ Score_____ Time_____

(1) 9 − 4		(26) 14 − 10		(51) 2 − 2		(76) 11 − 4	
(2) 1 − 1		(27) 10 − 6		(52) 15 − 6		(77) 6 − 6	
(3) 6 − 3		(28) 12 − 3		(53) 7 − 3		(78) 7 − 0	
(4) 12 − 10		(29) 10 − 10		(54) 15 − 7		(79) 11 − 10	
(5) 10 − 2		(30) 11 − 5		(55) 12 − 9		(80) 9 − 5	
(6) 17 − 9		(31) 4 − 0		(56) 7 − 7		(81) 11 − 1	
(7) 10 − 3		(32) 16 − 6		(57) 5 − 3		(82) 7 − 4	
(8) 9 − 1		(33) 14 − 8		(58) 16 − 8		(83) 12 − 7	
(9) 17 − 8		(34) 4 − 3		(59) 9 − 7		(84) 19 − 10	
(10) 6 − 5		(35) 9 − 6		(60) 14 − 9		(85) 13 − 4	
(11) 4 − 2		(36) 18 − 10		(61) 15 − 5		(86) 10 − 5	
(12) 19 − 9		(37) 9 − 0		(62) 13 − 8		(87) 13 − 3	
(13) 7 − 2		(38) 17 − 10		(63) 10 − 4		(88) 12 − 5	
(14) 5 − 4		(39) 8 − 7		(64) 11 − 8		(89) 15 − 10	
(15) 12 − 4		(40) 6 − 1		(65) 16 − 9		(90) 11 − 6	
(16) 15 − 8		(41) 11 − 7		(66) 2 − 1		(91) 11 − 2	
(17) 11 − 9		(42) 13 − 9		(67) 7 − 6		(92) 18 − 8	
(18) 10 − 7		(43) 5 − 2		(68) 3 − 0		(93) 8 − 4	
(19) 11 − 3		(44) 8 − 5		(69) 9 − 8		(94) 13 − 6	
(20) 14 − 7		(45) 13 − 5		(70) 18 − 9		(95) 8 − 1	
(21) 15 − 9		(46) 16 − 7		(71) 8 − 2		(96) 14 − 6	
(22) 12 − 8		(47) 14 − 4		(72) 13 − 7		(97) 6 − 4	
(23) 4 − 1		(48) 10 − 8		(73) 9 − 3		(98) 8 − 0	
(24) 8 − 3		(49) 16 − 10		(74) 14 − 5		(99) 4 − 4	
(25) 20 − 10		(50) 12 − 6		(75) 3 − 2		(100) 9 − 2	

84

Subtraction

(1) 9 – 4	5	(26) 14 –10	4	(51) 2 – 2	0	(76) 11 – 4	7
(2) 1 – 1	0	(27) 10 – 6	4	(52) 15 – 6	9	(77) 6 – 6	0
(3) 6 – 3	3	(28) 12 – 3	9	(53) 7 – 3	4	(78) 7 – 0	7
(4) 12 –10	2	(29) 10 –10	0	(54) 15 – 7	8	(79) 11 –10	1
(5) 10 – 2	8	(30) 11 – 5	6	(55) 12 – 9	3	(80) 9 – 5	4
(6) 17 – 9	8	(31) 4 – 0	4	(56) 7 – 7	0	(81) 11 – 1	10
(7) 10 – 3	7	(32) 16 – 6	10	(57) 5 – 3	2	(82) 7 – 4	3
(8) 9 – 1	8	(33) 14 – 8	6	(58) 16 – 8	8	(83) 12 – 7	5
(9) 17 – 8	9	(34) 4 – 3	1	(59) 9 – 7	2	(84) 19 –10	9
(10) 6 – 5	1	(35) 9 – 6	3	(60) 14 – 9	5	(85) 13 – 4	9
(11) 4 – 2	2	(36) 18 –10	8	(61) 15 – 5	10	(86) 10 – 5	5
(12) 19 – 9	10	(37) 9 – 0	9	(62) 13 – 8	5	(87) 13 – 3	10
(13) 7 – 2	5	(38) 17 –10	7	(63) 10 – 4	6	(88) 12 – 5	7
(14) 5 – 4	1	(39) 8 – 7	1	(64) 11 – 8	3	(89) 15 –10	5
(15) 12 – 4	8	(40) 6 – 1	5	(65) 16 – 9	7	(90) 11 – 6	5
(16) 15 – 8	7	(41) 11 – 7	4	(66) 2 – 1	1	(91) 11 – 2	9
(17) 11 – 9	2	(42) 13 – 9	4	(67) 7 – 6	1	(92) 18 – 8	10
(18) 10 – 7	3	(43) 5 – 2	3	(68) 3 – 0	3	(93) 8 – 4	4
(19) 11 – 3	8	(44) 8 – 5	3	(69) 9 – 8	1	(94) 13 – 6	7
(20) 14 – 7	7	(45) 13 – 5	8	(70) 18 – 9	9	(95) 8 – 1	7
(21) 15 – 9	6	(46) 16 – 7	9	(71) 8 – 2	6	(96) 14 – 6	8
(22) 12 – 8	4	(47) 14 – 4	10	(72) 13 – 7	6	(97) 6 – 4	2
(23) 4 – 1	3	(48) 10 – 8	2	(73) 9 – 3	6	(98) 8 – 0	8
(24) 8 – 3	5	(49) 16 –10	6	(74) 14 – 5	9	(99) 4 – 4	0
(25) 20 –10	10	(50) 12 – 6	6	(75) 3 – 2	1	(100) 9 – 2	7

Multiplication I

85

Name_____ Date_____ Score _____ Time _____

#	Problem		#	Problem		#	Problem		#	Problem	
(1)	5 x 3		(26)	5 x 4		(51)	6 x 3		(76)	9 x 2	
(2)	9 x 3		(27)	6 x 5		(52)	0 x 10		(77)	1 x 10	
(3)	8 x 9		(28)	9 x 8		(53)	0 x 2		(78)	8 x 8	
(4)	7 x 1		(29)	7 x 4		(54)	8 x 3		(79)	4 x 8	
(5)	9 x 6		(30)	3 x 7		(55)	3 x 6		(80)	3 x 2	
(6)	7 x 9		(31)	4 x 6		(56)	7 x 8		(81)	10 x 4	
(7)	10 x 3		(32)	4 x 3		(57)	10 x 2		(82)	6 x 9	
(8)	2 x 4		(33)	0 x 4		(58)	5 x 2		(83)	4 x 1	
(9)	8 x 5		(34)	6 x 6		(59)	8 x 10		(84)	9 x 5	
(10)	3 x 8		(35)	2 x 1		(60)	2 x 2		(85)	10 x 1	
(11)	6 x 10		(36)	4 x 2		(61)	8 x 4		(86)	1 x 6	
(12)	9 x 4		(37)	4 x 10		(62)	3 x 10		(87)	4 x 4	
(13)	0 x 1		(38)	1 x 2		(63)	6 x 1		(88)	10 x 10	
(14)	9 x 9		(39)	2 x 8		(64)	0 x 5		(89)	6 x 2	
(15)	4 x 7		(40)	8 x 7		(65)	3 x 1		(90)	3 x 9	
(16)	8 x 2		(41)	7 x 2		(66)	0 x 7		(91)	6 x 10	
(17)	1 x 9		(42)	7 x 6		(67)	9 x 1		(92)	3 x 8	
(18)	10 x 7		(43)	3 x 3		(68)	5 x 8		(93)	8 x 5	
(19)	1 x 1		(44)	0 x 8		(69)	2 x 3		(94)	2 x 4	
(20)	6 x 7		(45)	2 x 10		(70)	1 x 8		(95)	10 x 3	
(21)	7 x 10		(46)	7 x 3		(71)	10 x 6		(96)	7 x 9	
(22)	7 x 5		(47)	10 x 9		(72)	4 x 9		(97)	9 x 6	
(23)	5 x 9		(48)	2 x 7		(73)	2 x 9		(98)	7 x 1	
(24)	10 x 8		(49)	5 x 6		(74)	8 x 6		(99)	8 x 9	
(25)	5 x 5		(50)	3 x 5		(75)	3 x 4		(100)	4 x 5	

85

86

Multiplication I

(1)	5 x 3	15	(26)	5 x 4	20	(51)	6 x 3	18	(76)	9 x 2	18
(2)	9 x 3	27	(27)	6 x 5	30	(52)	0 x 10	0	(77)	1 x 10	10
(3)	8 x 9	72	(28)	9 x 8	72	(53)	0 x 2	0	(78)	8 x 8	64
(4)	7 x 1	7	(29)	7 x 4	28	(54)	8 x 3	24	(79)	4 x 8	32
(5)	9 x 6	54	(30)	3 x 7	21	(55)	3 x 6	18	(80)	3 x 2	6
(6)	7 x 9	63	(31)	4 x 6	24	(56)	7 x 8	56	(81)	10 x 4	40
(7)	10 x 3	30	(32)	4 x 3	12	(57)	10 x 2	20	(82)	6 x 9	54
(8)	2 x 4	8	(33)	0 x 4	0	(58)	5 x 2	10	(83)	4 x 1	4
(9)	8 x 5	40	(34)	6 x 6	36	(59)	8 x 10	80	(84)	9 x 5	45
(10)	3 x 8	24	(35)	2 x 1	2	(60)	2 x 2	4	(85)	10 x 1	10
(11)	6 x 10	60	(36)	4 x 2	8	(61)	8 x 4	32	(86)	1 x 6	6
(12)	9 x 4	36	(37)	4 x 10	40	(62)	3 x 10	30	(87)	4 x 4	16
(13)	0 x 1	0	(38)	1 x 2	2	(63)	6 x 1	6	(88)	10 x 10	100
(14)	9 x 9	81	(39)	2 x 8	16	(64)	0 x 5	0	(89)	6 x 2	12
(15)	4 x 7	28	(40)	8 x 7	56	(65)	3 x 1	3	(90)	3 x 9	27
(16)	8 x 2	16	(41)	7 x 2	14	(66)	0 x 7	0	(91)	6 x 10	60
(17)	1 x 9	9	(42)	7 x 6	42	(67)	9 x 1	9	(92)	3 x 8	24
(18)	10 x 7	70	(43)	3 x 3	9	(68)	5 x 8	40	(93)	8 x 5	40
(19)	1 x 1	1	(44)	0 x 8	0	(69)	2 x 3	6	(94)	2 x 4	8
(20)	6 x 7	42	(45)	2 x 10	20	(70)	1 x 8	8	(95)	10 x 3	30
(21)	7 x 10	70	(46)	7 x 3	21	(71)	10 x 6	60	(96)	7 x 9	63
(22)	7 x 5	35	(47)	10 x 9	90	(72)	4 x 9	36	(97)	9 x 6	54
(23)	5 x 9	45	(48)	2 x 7	14	(73)	2 x 9	18	(98)	7 x 1	7
(24)	10 x 8	80	(49)	5 x 6	30	(74)	8 x 6	48	(99)	8 x 9	72
(25)	5 x 5	25	(50)	3 x 5	15	(75)	3 x 4	12	(100)	4 x 5	20

Multiplication II

Name _____ Date _____ Score _____ Time _____

(1) 8 x 3		(26) 3 x 2		(51) 2 x 6		(76) 6 x 9	
(2) 0 x 10		(27) 6 x 2		(52) 9 x 10		(77) 10 x 8	
(3) 0 x 1		(28) 9 x 7		(53) 1 x 5		(78) 7 x 5	
(4) 6 x 10		(29) 6 x 4		(54) 10 x 5		(79) 6 x 7	
(5) 8 x 5		(30) 2 x 5		(55) 4 x 10		(80) 10 x 7	
(6) 10 x 3		(31) 1 x 3		(56) 2 x 1		(81) 8 x 2	
(7) 9 x 6		(32) 5 x 10		(57) 0 x 4		(82) 9 x 9	
(8) 8 x 9		(33) 2 x 8		(58) 10 x 6		(83) 6 x 5	
(9) 5 x 3		(34) 7 x 2		(59) 2 x 9		(84) 7 x 4	
(10) 4 x 4		(35) 3 x 3		(60) 3 x 4		(85) 4 x 6	
(11) 10 x 1		(36) 2 x 10		(61) 3 x 10		(86) 5 x 8	
(12) 4 x 1		(37) 10 x 9		(62) 2 x 2		(87) 9 x 2	
(13) 10 x 4		(38) 5 x 6		(63) 5 x 2		(88) 4 x 8	
(14) 5 x 5		(39) 1 x 4		(64) 7 x 8		(89) 3 x 9	
(15) 5 x 9		(40) 3 x 6		(65) 0 x 2		(90) 8 x 1	
(16) 7 x 10		(41) 10 x 2		(66) 6 x 3		(91) 5 x 7	
(17) 1 x 1		(42) 8 x 10		(67) 9 x 4		(92) 7 x 7	
(18) 1 x 9		(43) 8 x 4		(68) 3 x 8		(93) 6 x 8	
(19) 4 x 7		(44) 6 x 1		(69) 2 x 4		(94) 4 x 5	
(20) 5 x 4		(45) 8 x 6		(70) 7 x 9		(95) 8 x 7	
(21) 9 x 8		(46) 4 x 9		(71) 7 x 1		(96) 7 x 6	
(22) 3 x 7		(47) 1 x 8		(72) 9 x 3		(97) 0 x 8	
(23) 2 x 3		(48) 6 x 6		(73) 10 x 10		(98) 7 x 3	
(24) 3 x 1		(49) 4 x 2		(74) 1 x 6		(99) 2 x 7	
(25) 8 x 8		(50) 1 x 2		(75) 9 x 5		(100) 3 x 5	

Multiplication II

#			#			#			#			#		
(1)	8 x 3	24	(26)	3 x 2	6	(51)	2 x 6	12	(76)	6 x 9	54
(2)	0 x 10	0	(27)	6 x 2	12	(52)	9 x 10	90	(77)	10 x 8	80
(3)	0 x 1	0	(28)	9 x 7	63	(53)	1 x 5	5	(78)	7 x 5	35
(4)	6 x 10	60	(29)	6 x 4	24	(54)	10 x 5	50	(79)	6 x 7	42
(5)	8 x 5	40	(30)	2 x 5	10	(55)	4 x 10	40	(80)	10 x 7	70
(6)	10 x 3	30	(31)	1 x 3	3	(56)	2 x 1	2	(81)	8 x 2	16
(7)	9 x 6	54	(32)	5 x 10	50	(57)	0 x 4	0	(82)	9 x 9	81
(8)	8 x 9	72	(33)	2 x 8	16	(58)	10 x 6	60	(83)	6 x 5	30
(9)	5 x 3	15	(34)	7 x 2	14	(59)	2 x 9	18	(84)	7 x 4	28
(10)	4 x 4	16	(35)	3 x 3	9	(60)	3 x 4	12	(85)	4 x 6	24
(11)	10 x 1	10	(36)	2 x 10	20	(61)	3 x 10	30	(86)	5 x 8	40
(12)	4 x 1	4	(37)	10 x 9	90	(62)	2 x 2	4	(87)	9 x 2	18
(13)	10 x 4	40	(38)	5 x 6	30	(63)	5 x 2	10	(88)	4 x 8	32
(14)	5 x 5	25	(39)	1 x 4	4	(64)	7 x 8	56	(89)	3 x 9	27
(15)	5 x 9	45	(40)	3 x 6	18	(65)	0 x 2	0	(90)	8 x 1	8
(16)	7 x 10	70	(41)	10 x 2	20	(66)	6 x 3	18	(91)	5 x 7	35
(17)	1 x 1	1	(42)	8 x 10	80	(67)	9 x 4	36	(92)	7 x 7	49
(18)	1 x 9	9	(43)	8 x 4	32	(68)	3 x 8	24	(93)	6 x 8	48
(19)	4 x 7	28	(44)	6 x 1	6	(69)	2 x 4	8	(94)	4 x 5	20
(20)	5 x 4	20	(45)	8 x 6	48	(70)	7 x 9	63	(95)	8 x 7	56
(21)	9 x 8	72	(46)	4 x 9	36	(71)	7 x 1	7	(96)	7 x 6	42
(22)	3 x 7	21	(47)	1 x 8	8	(72)	9 x 3	27	(97)	0 x 8	0
(23)	2 x 3	6	(48)	6 x 6	36	(73)	10 x 10	100	(98)	7 x 3	21
(24)	3 x 1	3	(49)	4 x 2	8	(74)	1 x 6	6	(99)	2 x 7	14
(25)	8 x 8	64	(50)	1 x 2	2	(75)	9 x 5	45	(100)	3 x 5	15

These and other materials are available from Instructional Resources Co.

	Quantity	Extended Cost

Materials to Help Students Learn Math

*Addition Facts in Five Minutes a Day** Item 056 $11.95   _____ _____

*Subtraction Facts in Five Minutes a Day** Item 064 $11.95   _____ _____

*Multiplication Facts in Five Minutes a Day** Item 072 $13.95   _____ _____

*Division Facts in Five Minutes a Day** Item 080 $11.95   _____ _____

*Casting 9's: A Quick Check for Math Computation** Item 099 $8.00   _____ _____

Materials to Help Students Learn Reference

Facts Plus: An Almanac of Essential Information ©1999, 250 pages,
   ISBN 1-879478-19-6 Item 196 $15.95   _____ _____
   Item 19D, $11.96 each for orders of 25 or more almanacs.   _____ _____

*Facts Plus Activity Book** ©1995, 182 pages, ISBN 1-879478-11-0
   Item 110 $19.95   _____ _____

Encyclopedia Activity for use with *The World Book Encyclopedia**
   ©1996, Item 129 $9.50   _____ _____

Materials to Help Students Learn Spelling

*Spelling Plus: 1000 Words Toward Spelling Success** Item 048 $19.95   _____ _____

Dictation Resource Book for use with *Spelling Plus* Item 137 $12.95   _____ _____

*Homophones Resource Book** Item 145 $15.95   _____ _____

*Personal Dictionary** Item 153 $3.50   _____ _____

Spell Well: A One-Year Review for Older Students Item 042 $11.95   _____ _____

Materials to Help Students Learn Handwriting

*Manuscript Handwriting Masters** Item 161 $8.50   _____ _____

*Cursive Handwriting Masters** Item 17X $10.50   _____ _____

*contains reproducible masters

Total   _____

Shipping ($3.20 for first book,
.50 for each additional book)   _____

Total enclosed   _____

Instructional Resources Company
3235 Garland
Wheat Ridge, CO 80033
    or
Instructional Resources Company
P.O. Box 111704
Anchorage, AK 99511-1704
907-345-6689 (Phone / FAX)

Name _____

Organization _____

Address _____

_____

Phone number _____

❏ Teacher   ❏ Home School ❏ Library   ❏ Other